Gravity, Strings and Particles

Maurizio Gasperini

Gravity, Strings and Particles

A Journey Into the Unknown

 Springer

Maurizio Gasperini
Dipartimento di Fisica
Università di Bari
Bari, Italy

ISBN 978-3-319-00598-0   ISBN 978-3-319-00599-7 (eBook)
DOI 10.1007/978-3-319-00599-7
Springer Cham Heidelberg New York Dordrecht London

Library of Congress Control Number: 2014937666

Printed on acid-free paper

Springer is part of Springer Science+Business Media (www.springer.com)

To my Mother
with my deepest gratitude

Preface

This book grew out of conversations I had with a good friend of mine (Pier Paolo Casalboni, also known as "*Slim*"). During the summer vacations, while tanning on the beach in Cesenatico, he often asks me to tell him the latest news and the most peculiar ideas concerning my work as theoretical physicist.

In this book I will talk of physics addressing to readers who do not necessarily have a specific background in this field, but are nevertheless interested in discovering the novelty, the originality, and the possible weird implications of some amazing ideas used by modern physics of fundamental interactions. I will avoid introducing mathematical expressions as much as possible, trying to convey ideas rather than explaining formulas. Also, I will leave aside the cautious attitude typical of the academic style, following sometimes my excitement and my personal feelings concerning the topics under discussion.

We can say that this is a book of popular science, but of a rather unconventional type, as the emphasis is not only on what is known but also—and mainly—on what is still unknown. Indeed, many parts of the book are devoted to introduce and illustrate fundamental theoretical models and results which are potentially highly relevant to a deeper understanding of Nature, but still waiting to be directly confirmed (or disproved) by experimental observations. From this point of view the book may be of some interest also to professional physicists, working or not in the field of fundamental interactions.

I should explain, finally, the reason why the book is focused on the three topics mentioned in the title: gravity, strings, and particles. Why these three topics? What brings them together, selecting them among many other important issues of modern physical research?

It is known that there are many important links among them: for instance, as we shall see, the fact that a unified description of all elementary matter particles and all forces, including gravity, can be consistently achieved only within a model based on strings.

However, my choice is mainly motivated by the widespread belief that only a joint study of high-energy models of gravity, strings, and particles may help us to shed light on what seems to be (to me, at least) one of the biggest and most fascinating mysteries of modern science: besides time and three spatial dimensions, are there other dimensions in our Universe? If yes, how many are they?

Cesena, Italy   Maurizio Gasperini
February 2013

Notations

For a maximum simplification of the (very few) equations presented in this book I will always adopt the so-called natural system of units, where both the light velocity c and the Planck constant $\hbar$ are set equal to one.

With this choice of units mass and energy have the same physical dimensions, energy has dimensions of the inverse of a length, and energy density has dimensions of an inverse length to the fourth power.

We will often use, as a typical reference distance, the Planck length L_P defined (in the above units) by $L_P = \sqrt{G}$, where G is Newton's gravitational constant; we will also use, as a typical reference energy, the Planck mass M_P defined by $M_P = 1/L_P$.

We will usually express distances in centimeters (abbreviated as cm); energies in electron volts (abbreviated as eV), or billions of electron volts (abbreviated as GeV), or thousands of GeV (abbreviated as TeV). Sometimes we will use for the distances also the light year, roughly equivalent to 0.9×10^{18} cm. Finally, we will express temperatures in Kelvin degrees, remembering that one Kelvin degree (with the Boltzmann constant set equal to one) corresponds to 8.6×10^{-5} eV.

In the above units the Planck length is given by:

$$L_P \simeq 1.61 \times 10^{-33} \text{ cm},$$

the Planck mass is given by:

$$M_P \simeq 1.22 \times 10^{19} \text{ GeV},$$

and the Hubble radius L_H, which controls the size of the portion of space directly accessible to our observation, is presently given by:

$$L_H \simeq 1.28 \times 10^{28} \text{ cm}.$$

Other scales of energy and distance, possibly relevant to the topics of this book, will be introduced and discussed whenever necessary.

Contents

Chapter 1
Prologue: Inside the Energy Walls of Our "Cradle"

We often say that the physics of "small distances" is equivalent to the physics of "high energies." This is indeed true, as a direct consequence of the celebrated Heisenberg's principle (or uncertainty principle) stating that, in order to explore (and measure) smaller and smaller distances, we need probes with higher and higher momenta, namely with larger and larger kinetic energies. According to the uncertainty principle, in particular, the required energy E turns out to be inversely proportional to the considered distance d, so that E tends to infinity when the distance d goes to zero.

Even in the case of very large distances, however, we are unavoidably lead to the high-energy regime. This basically occurs for two reasons: one reason, of accidental type, is related to the expansion of our Universe; the other reason, of more fundamental nature, is related to the fact that all information and signals (of all types) are characterized by a finite speed of propagation.

According to this second (important) property of Nature, in fact, looking "far away in space" also means looking "back in time," because the signals we receive from more and more distant sources have been emitted at increasingly remote epochs. If a galaxy is millions of light-years away from Earth, for instance, its light has been traveling for millions of years to get to us, and the information it can provide is referred to the epoch when the light left the galaxy—namely, to millions of years ago.[1]

Because of the expansion of our Universe, on the other hand, looking back in time implies considering epochs in which matter and radiation were concentrated in increasingly smaller volumes of space, so that the temperature and the kinetic energy of their elementary components were higher and higher. Hence, the more remote is the signal which reaches us, the greater is the energy scale corresponding to the emission epoch.

---

[1] The famous *Andromeda Galaxy*, whose picture is also used as a desktop background in recent versions of Mac computers, is one of the nearest galaxies, and is approximately 2.5 million light-years away from Earth (corresponding to a distance of about 2.4×10^{19} km).

M. Gasperini, *Gravity, Strings and Particles*, DOI 10.1007/978-3-319-00599-7_1,
© Springer International Publishing Switzerland 2014

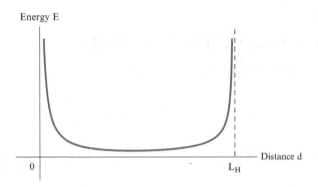

Fig. 1.1 The energy scale E as a function of the corresponding distance scale d. The physically accessible range of distances seems to be bounded by two walls of infinitely high energy

It follows that our observations cannot be extended back in time (and out in space) at our will: beyond a given epoch, for instance, the Universe is so dense as to be no longer transparent to the electromagnetic radiation[2] (the emitted light is immediately reabsorbed, hence it cannot get to us today and bring us information about those eras).

We might consider different types of radiation (for instance, gravitational waves) which are more penetrating than light, and can reach us from more remote eras. Even proceeding in this way, however, standard cosmology tells us that we *must* encounter, at a given time and to a given distance, an impassable barrier due to the so-called "initial singularity:" the famous Big Bang.

The Big bang singularity, which marks the beginning of the cosmological expansion, and which is characterized by an arbitrarily high-energy scale, is not infinitely remote in time (and distant in space): it is localized at an epoch that approximately dates back to 14 billion years ago, and that corresponds to a spatial distance of the order of the so-called "Hubble radius," L_H. Such a distance is time dependent, and its present value is just about 14 billion light-years. For spatial distances approaching L_H the corresponding energy scale tends to infinity.

In order to summarize the previous discussion, and synthesize our findings, we can produce a (empirical) plot of the energy scale E as a function of the distance d. We obtain in this way a curve like the one reported in Fig. 1.1, characterized by an unbounded growth of the energy in the limit of both very small distances ($d \to 0$) and very large distances ($d \to L_H$), approaching the Hubble radius.

Such a behavior of E seems to keep our observation capability confined within a limited range, bounded by two physically insourmountable walls. In fact, an

---

[2]This occurs when the radiation reaches a temperature that is about a 1,000 times larger than the current one: more precisely, a temperature of 2,973 K. Such a temperature is reached at the so-called "decoupling epoch," see for instance the textbooks by Durrer [1], Weinberg [2], Gasperini [3] (in Italian).

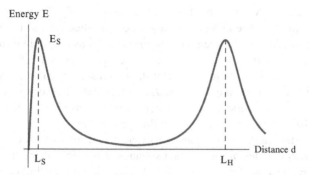

Fig. 1.2 The energy scale E as a function of the corresponding distance scale d, including the energy bounds suggested by string theory. The physically accessible distances now range from zero to arbitrarily high values

infinitely high energy would seem to be required to get access to arbitrarily small and/or arbitrarily large distances, just as if Nature had prepared for us a "cradle" from which we cannot escape.

As every cradle, however, also the "energy cradle" we are considering might prove effective to confine and protect a "newborn" physical science, becoming however inadequate, and no longer impassable, with the growth and the ripening of our scientific knowledge. There are indeed recent developments in theoretical physics, to be illustrated in the following chapters, suggesting that the energy walls of Fig. 1.1 might be "smoothed out"—at both large and small distances—and replaced by barriers of very high but *finite* energy.

Anticipating some results, and considering first the "cosmological" barrier associated with the Big Bang, we may recall that the modern string theory allows to formulate models of the Universe in which the initial singularity is replaced by a transition phase—the so-called "string phase"—with typical values of temperature and density much higher than those of ordinary macroscopic matter, but *not infinite*. In that case the energy scale E is no longer divergent as the distance approaches L_H, but it is limited to a maximum value E_S (determined by string theory). At larger distances the energy goes back to the decreasing regime, allowing (at least in principle) the observation of spatial distances (and time intervals) of arbitrarily large extension (see Fig. 1.2).

We may expect a similar change also for the energy barrier located at small distances. In fact, the scale of maximum energy E_S is inversely proportional to a distance scale which we shall call L_S, and which is typical of the theories of strings and extended objects in their quantum version. Below that distance, which represents the minimal length of the quantized string (or extended object), we may expect that the uncertainty relation may acquire corrections able to remove the infinite amount of energy fluctuations associated with the presence of infinitely small distances, so as to fix a maximum energy scale in correspondence of the string length L_S.

The outcome of the above modifications is qualitatively illustrated in Fig. 1.2, showing how the two energy barriers might be smoothed around the two critical distances L_S and L_H, as a consequence of the corrections induced by string theory.

Given that the above figure is not in scale and does not reflect the actual proportions, it is appropriate to stress that the two distance scales L_S and L_H are tremendously different from each other: L_S corresponds to a very small length, of the order of 10^{-32} cm, while L_H is extremely large and (as already mentioned) is of the order of 10^{28} cm (approximately 14 billion light-years).

Also, the peak value of the energy barrier, E_S, is enormous with respect to the typical energy scales of nuclear and subnuclear physics. String theory, in fact, suggests for E_S a value of the order of 10^{15} TeV: this is a million billion times larger than the maximum energy presently reached by the world's biggest accelerator "Large Hadron Collider" (LHC), operating at the CERN laboratories near Geneva.

The two barriers we are considering are thus of finite but very large height and are located at an enormous distance from each other. Which other worlds, and what new natural phenomena, are waiting for us beyond those barriers regarded as impassable by the physics of the last century?

We are a bit intrigued and a bit intimidated, just like an infant raising his head for the first time to look over the walls of his cradle.

Chapter 2
Gravity at Small Distances

Among all fundamental forces of Nature, gravity is probably the one we think we know better—if only because it is the one which has always influenced our experience and our way of life, since the beginning of the human history.

At the high school we still learn Newton's law of universal gravitation: given two masses, they attract each other with a force which is inversely proportional to the square of their distance. According to such law, if we halve the distance the force becomes four times stronger. If the distance is reduced to one quarter, the force becomes 16 times stronger. And so on. But what happens if we go to smaller and smaller distances? Are we sure that the gravitational force keeps behaving as predicted by Newton?

I should insert, at this point, an important remark. The small distances I am talking about, for the moment, are not "so small" to require the application of the principles of quantum physics. If we enter the regime where we need to "quantize" the gravitational interaction we know, indeed, that there will be corrections due to the production of virtual particles,[1] and that the classical gravitational laws will be unavoidably—and maybe drastically—modified. I don't want to consider such corrections, for the moment, and so let us confine ourselves to a range of distances where classical physics is still valid.

I should also stress that, even restricting ourselves to the classical context, Newton's theory only provides an approximate and incomplete description of the gravitational interaction. The correct gravitational model—according to modern science—is given by Einstein's theory of general relativity, representing gravity as a geometric consequence of the space–time curvature. However, in the limit in which the sources are static, the gravitational field is sufficiently weak, and the spatial curvature is so small to be negligible, even the Einstein theory predicts that the force between two point-like masses should follow the inverse-square law, exactly as predicted by Newton.

---

[1] For those who are expert of physics it will be clear that I am referring to the quantum "loop" corrections, described by a series of Feynman graphs of increasing accuracy and complexity.

M. Gasperini, *Gravity, Strings and Particles*, DOI 10.1007/978-3-319-00599-7_2,
© Springer International Publishing Switzerland 2014

Hence, the question we asked previously is well posed. Up to which distance can we trust the classical law of Newtonian gravity? The answer can only be provided by direct experimental tests, performed at smaller and smaller distances, up to the maximum limits allowed by current technology.

Experiments testing the inverse-square law of Newtonian gravity have been—and currently are—performed with ever increasing precision. Many possible types of corrections have been considered.

For instance the possibility that, at small enough distances, the gravitational force is inversely proportional *not* to the square but to the cube, or to the fourth power, or to some other power of the distance. Or the possibility that the force is exponentially decreasing with the distance. And even the possibility that Newton's gravitational constant—the famous constant G—is not a universal parameter, and that its precise value may change with the distance.

All these possible modifications of Newton's law have been tested with modern high-precision instruments able to measure the gravitational force at small distances[2]: torsion balances, torsion pendulums, high-frequency and low-frequency torsion oscillators, and the so-called *microcantilevers* (tiny silicon slivers arranged as small trampolines and acting as microscopic vibrators).

No experiment has been able, to date, to detect any violation of the inverse-square law predicted by Newton. Assuming that such violations are controlled by a gravitational constant which is always the same at all distances, then present experimental tests tell us that possible modifications of Newton's law are only allowed on submillimetric scales of distance: more precisely, only below a distance of about 2 dmm, i.e., 2×10^{-2} cm (we may expect that such a limit will be soon extended down to 10^{-3} cm). At larger distances (of the order, say, of the centimeter), possible modifications are allowed only if they are characterized by a gravitational constant—namely, by an effective intensity—which is at least 1,000 times smaller than the Newton constant G.

The above experimental results are rather surprising, and not because they provide a rather accurate verification of Newton's law of gravitational attraction. On the contrary, just for the opposite reason: in fact, Newton's law turns out to be confirmed only up to distances which are little below the millimeter scale, thus leaving plenty of room for possible changes.

We can observe, indeed, that the general relativistic corrections to Newton's law start to play a crucial role only at distances of the order of the so-called Schwarzschild radius. This means—for the typical masses, of the order of 1 kg, used in the experiments—only at distances of about 10^{-25} cm. The corrections possibly due to quantum effects, on the other hand, become important at even much smaller distances, of the order of the Planck length L_P (which is about 10^{-33} cm). Between these scales of distances and those explored by present experimental tests there is clearly a huge "vacuum" of information.

---

[2]The interested reader can find a description of such experiments in the review paper by Adelberger et al. [4].

· Hidden in the unexplored "desert" of the intermediate distances, important modifications of Newton's law of gravity are possibly waiting to be tested by more accurate experiments. Such modifications, once discovered (or directly disproved) could help us to achieve a better understanding of Nature and of its fundamental interactions.

There are indeed many theoretical models predicting the possibility—and even the need—of introducing corrections to Newton's law at small distances. As we shall see in the rest of this chapter, such corrections are in principle of three types: corrections due to (1) new forces of Nature; or (2) new intrinsic properties of the gravitational interaction; or (3) new (and additional) spatial dimensions.

2.1 New Fundamental Forces of Nature?

In the second half of the 1980, when I was a young researcher in the staff of the Department of Theoretical Physics at the University of Turin, I remember that there was quite a stir for a paper written by a group of US physicists,[3] claiming the experimental discovery of a violation of the equivalence principle.

Such a celebrated principle, representing one of the conceptual pillars of Einstein's theory of general relativity, states that the gravity acts "universally"— i.e., with the same coupling strength—on all forms of matter and energy. Unlike the electromagnetic force, which distinguishes between positive, negative, and null charges, gravity always induces the same acceleration on all bodies, and there is no body which is "neutral" under the action of the gravitational force.[4]

It is just because of this property that all gravitational effects can always be eliminated, on a sufficiently small portion of space and for small enough time intervals. We may recall, in this respect, an effect probably seen by many of us, while watching on TV movies the astronauts orbiting around the Earth: the astronauts are freely floating inside their cockpit together with other lighter objects (a notebook, a pencil), just as if gravity would be totally absent in that place, for all bodies, quite irrespective of their mass.

Now, the quoted paper by Fisbach et al. [5] reported a reanalysis of a famous experimental test of the equivalence principle performed by Eötvös, Pekár, and Fekete at the beginning of the nineteenth century. The conclusion of such reanalysis was that the gravitational force between metallic bodies of different types (in particular, between copper and aluminum bodies, and between copper and lead bodies) was characterized by different intensities, just as if different material had different "gravitational charges" (or different effective values of the gravitational coupling constant). An explicit violation of the universality principle! The effective differences, in particular, were estimated to be little below the 1 % level.

---

[3] It was a paper by Fischbach et al. [5].

[4] At least, to the best of our present knowledge.

It should be immediately stressed that these results have been subsequently disproved by further, more accurate, experimental analyses, and that no deviation from the universality of the gravitational interaction has been found and confirmed to date. Nevertheless, the claim of a presumed violation of the equivalence principle had the virtue of triggering a spate of works and research papers, discussing the possible interpretations and implications of such violation in the context of theoretical models of gravity and of the other fundamental interactions.

The question (interesting in itself) we may ask, in fact, is the following: how to explain, and theoretically describe, the fact that different materials may "feel" and "react" to gravity differently? Also, and above all, how to introduce such an effect without contradicting the well-known results of Newton's and Einstein's theories which—after all—are perfectly consistent with the observed gravitational phenomena at large distances? (consider, for instance, the planetary motion, the precession of planetary orbits, and so on).

A simple answer can be given, in principle, by the theoretical models predicting that the total effective gravitational force contains the contributions of two (or more) components. One component has an infinite range (hence it is dominant at large distances) and acts universally, with a strength controlled by Newton constant G, on the total mass of all bodies. The other component, instead, is a short-range force directly acting on the atomic components (protons, neutrons, electrons) of the given bodies, with a coupling strength similar (but not necessarily identical) to the gravitational constant G.

Such a new component is also called *fifth force,* in order to stress its close relationship with the other four fundamental forces of Nature: electromagnetic, gravitational, weak, and strong nuclear forces. Given its short-range behavior, the fifth force may be detectable in laboratory experiments, but it is expected to be invisible at large enough distances (for instance, it should have no effect at all on the planetary motion and on the solar system dynamics). Also, since it is directly coupled to the atomic components, it may act differently on macroscopic bodies with different chemical composition (like, for instance, the copper, aluminum, lead bodies we mentioned earlier): hence, it may produce a total effective gravitational force which turns out to be *composition dependent.*

Are there fundamental theoretical models, not invented ad hoc to predict violations of the equivalence principle, which include the possible presence of a fifth force?

The answer (perhaps surprising) is yes. In particular, there are models of "supergravity" and "superstrings" (which will be dealt with in detail later) where the presence of small violations of the equivalence principle (and of other types of anomalies in the gravitational force), at small distances, is not only possible but also—under certain conditions—unavoidable.

In these models the fifth force may be represented by a field of vector type, transmitted by a particle called "graviphoton," or by a field of scalar type, transmitted by a particle called "dilaton." These two possibilities will be briefly illustrated in the following section.

2.1.1 The Graviphoton and the Dilaton

Let us start with the graviphoton, a particle predicted by the so-called models of "extended supersymmetry."[5]

We should recall, first of all, that a physical system is called "supersymmetric" if it contains the same number of bosonic and fermionic components, and if it is invariant under the exchange of its bosonic components with the fermionic ones. Bosons are particles (like photons, for instance) which have integer spin (i.e., integer *intrinsic* angular momentum) and obey the rules of Bose–Einstein statistics; fermions are particles (like electrons, for instance) which have half-integer spin and obey the rules of Fermi–Dirac statistics.[6]

Ordinary physical systems are usually invariant under transformations exchanging, separately, bosons among themselves and fermions among themselves. A supersymmetric system is thus a somewhat exceptional physical system. This is confirmed by the fact that, even including all presently known particles (and there are many!) and considering all their possible combinations, there is no way to build up a system satisfying the required properties of supersymmetry.

Supersymmetry, on the other hand, seems to be a needful—or, at least, a very useful—ingredient for the solution of many formal problems that unavoidably arise when attempting a unified description of all fundamental forces and all elementary components of matter. Hence, to achieve the goal of a successful unified theory, various theoretical models have postulated the existence of new particles which are still to be discovered,[7] but which possess the right properties to allow the construction of a supersymmetric description of Nature.

A comment on this procedure, at this point, is in order. Postulating the existence of new particles, to the only purpose of implementing a required symmetry and improving the formal properties of a theory, might seem to be a rather daring way to proceed. However, it is just that way that, in the past, we have made important discoveries in the physics of fundamental interactions. Suffice it to recall, in this respect, the discovery of the so-called vector bosons Z and W (the force carriers of the electro-weak interactions), theoretically postulated in the 1960 following a symmetry principle (the "gauge" invariance of those interactions), and experimentally detected only after several years, in particular thanks to the results

---

[5]See for instance the paper by Barbieri and Cecotti [6].

[6]In the case of the Bose–Einstein statistics there is no limit to the number of particles that can occupy the same quantum state. In the case of the Fermi–Dirac statistic, on the contrary, two or more particles cannot occupy the same quantum state (according to the so-called Pauli exclusion principle).

[7]The search of supersymmetric particles is one of the main goals of the experiments performed using the powerful accelerator LHC at the CERN laboratories in Geneva. At the time of writing (March 2013), however, no positive result has yet been reported.

obtained in 1983 by the particle accelerator SPS at the CERN laboratories in Geneva.[8]

Coming back to the topic of interest, we should stress that to obtain a super-symmetric model any particle must be associated to an appropriate "twin" particle or—as commonly said—to an appropriate supersymmetric "partner." For instance, if we want to include in our model the photon, which is the carrier of the electromagnetic interactions and is a boson particle with spin quantum number equal to 1, we must also introduce the so-called photino, which has similar interaction properties but is a fermion with spin 1/2. If we want to include the graviton, which is the carrier of the gravitational interaction and is a boson particle with spin number equal to 2, we must also introduce the "gravitino," which has similar interaction properties but is a fermion with spin 3/2. And so on.

For a mathematically consistent model, however, adding to each particle the corresponding supersymmetric partner might be not enough. In particular, if we are attempting to provide a unified description of all interactions, including all of them in the same supersymmetric schemes, we must introduce also additional particles whose role, in a sense, in that of connecting the various interactions among themselves.

All the particles present in the model can then be combined and classified in groups, called "multiplets," containing elementary components of the model participating in the same interactions. The gravitational multiplet, in particular, contains not only the graviton and its (already mentioned) fermionic partner, the gravitino, but also other boson particles like the graviphoton and the graviscalar.

Let us focus on the graviphoton, which is the one we are interested in for a discussion of the fifth force. It is a particle similar to the photon, since it is a boson with spin equal to 1. It differs from the photon, however, in three important aspects.

First of all it is massive, and thus propagates at a speed lower than the speed of light. Second, it is coupled not to the usual electric charge but to the so-called "baryonic charge," which is carried by the heavy particles (protons and neutrons) living inside the atomic nuclei.[9] Third, it mediates a force which is much weaker than the electromagnetic one, and which has an intensity similar—but not necessarily identical—to that of the gravitational force.

Like the electromagnetic force, however, also the force mediated by the graviphoton is of vector type, and thus it is attractive between baryonic charges of opposite sign and repulsive between baryonic charges of the same sign. This means, in particular, that the two protons (or two neutrons) will repel each other under the

---

[8]The existence of these particles was theoretically predicted by S. Glashow, S. Weinberg, and A. Salam. The final experimental confirmation is due to C. Rubbia and S. van de Meer. All these physicists have been subsequently awarded the Nobel Prize in Physics.

[9]More precisely, the graviphoton is coupled to the so-called hypercharge, defined by adding to the baryonic charge other elementary charges characterizing the different species of *quarks* (which are the elementary components of protons and neutrons). Such additional charges, however, are vanishing for the typical atomic nuclei of ordinary matter.

action of the graviphoton, while protons and antiprotons (as well as neutrons and antineutrons) will attract each other.

Given that ordinary matter does not contain antiparticles, the effect of the graviphoton, in that case, is to produce a repulsive force which must be added to the usual gravitational interaction, thus weakening the resulting overall attraction. The total effective force, in addition, acquires a dependence on the chemical composition of the bodies we are considering, because different materials have different number of protons and neutrons in their nuclei, different baryonic charges, and are thus differently affected by the graviphoton interaction.

If the graviphoton exists, why its effects have not been observed so far? Maybe because the mass of the graviphoton is very large and—as a consequence—the force mediated by the graviphoton has a range too small to fall within the scales of distance explored by the current gravitational experiments.

The range of a force, in fact, is inversely proportional to the mass of the carrier particle. The graviphoton mass, on the other hand, should be roughly determined by the energy scale at which the supersymmetry invariance between bosons and fermions breaks down.[10]

Since we are not observing any evidence of supersymmetry up to the presently accessible energy scales, which are of the order of the TeV (i.e., about 1,000 times the mass of the proton), it follows that supersymmetry stays broken up to scales larger (or at least equal to) the TeV scale, and this implies, for the graviphoton, a mass larger than (or at least of the same order as) this scale. The range of a force mediated by a TeV-mass particle is about 10^{-16} cm, which is indeed a distance by far smaller than those explored by the current measurements of the gravitational force (see the discussion at the beginning of this chapter).

Alternatively, we might speculate that the effects of the graviphoton have not yet been observed—even if its mass is very light, and the range of the force is correspondingly large—simply because its coupling to the baryonic charge is too weak to produce detectable consequences. This is, in fact, what predicted by models formulated in space–times with many dimensions. In any case, since the present gravitational experiments are not sensitive enough to detect graviphotons, we should ask ourselves if there are other possible methods of direct (or indirect) detection.

The answer is positive, at least in principle, as the graviphoton may interact with the photon, i.e., with the particle mediating the electromagnetic interaction: hence, besides modifying Newton's equations for the gravitational field, the graviphoton also modifies Maxwell's equations for the electromagnetic field itself.

---

[10]For a consistent supersymmetric model, in fact, all particles belonging to the same multiplet should be characterized by the same mass. Hence, in the exact supersymmetric regime, the mass of the graviphoton should be the same as the graviton mass (which is zero). When supersymmetry is broken, on the contrary, a mass difference is generated even among the particles of the same multiplet.

Thanks to the photon–graviphoton interaction, in particular, new interesting electromagnetic effects become possible.[11] We find, for instance, that a macroscopic body containing a net baryonic charge (which is roughly proportional to the total number of protons and neutrons contained inside the body) can be the source of an electric field even if the body is neutral! (namely, if its total electric charge is zero). The resulting electric field, however, is a short-range field, hence it tends to disappear at the level of macroscopic distances if—as discussed before—the mass of the graviphoton is sufficiently large.

Such an additional, short range field is clearly present even if the body is electrically charged. In that case the total produced electric field has a behavior which deviates from the well-known Coulomb's law, just as if the photon has acquired an effective mass (very small, but different from zero). The experimental tests of Coulomb's law—just like the tests of Newton's law of gravity—then give us direct information on the range of the graviphoton force, and on the strength of its coupling to photons.

Such a coupling, if it exists, turns out to be very small: the ratio between the intensities of the graviphoton and photon components of the total electric field, in fact, has to be smaller than about one part in a million, to avoid clashing with the present experimental results. In addition, it is important to stress that the electromagnetic corrections induced by a possible photon–graviphoton coupling—unlike the gravitational corrections induced by the baryon–graviphoton coupling—are (in first approximation) *independent* of the composition of the charged body (hence, in this sense, are of universal type).

We should recall, finally, another interesting consequence of the graviphoton interactions occurring even in vacuum, namely in the absence of charges and currents (of both electric and baryonic type).

An electromagnetic wave, freely propagating in vacuum with a given constant momentum, contains in general both the photon and the graviphoton component describing, respectively, massless and massive radiation. These two components are characterized by different energies, and thus by different frequencies: as a consequence they interfere, and their interference necessarily induces oscillations in the total intensity of the wave (for an effect of "mixing," or mutual conversion, of the two components, very similar to the effect responsible for the well-known neutrino oscillations).

This tiny oscillation effect, which is very difficult to be detected through direct laboratory experiments, could nevertheless produce important consequences in an astrophysical and cosmological context. In particular, it could be at the ground of the mechanism producing the cosmic magnetic fields observed on intergalactic scales,[12] whose origin still remains mysterious.

---

[11] See for instance an old paper I wrote in 1989 [7].

[12] I have discussed this possibility in a paper written in 2001 [8].

Let us now consider the possibility that the fifth force is mediated by a particle of scalar type.[13]

There are many theoretical models, in fact, predicting for the graviton a scalar partner: the (already mentioned) models of extended supersymmetry, for instance, require that the graviton is paired with the so-called graviscalar particle. However, there is no need of introducing supersymmetry in order to formulate interesting, and formally consistent, models of gravity including a scalar component: a popular example is the so-called Brans–Dicke model, a scalar–tensor model of gravity proposed half a century ago, and still valid.

Here we will concentrate, in particular, on the case of the so-called dilaton, the scalar partner of the graviton required by string theory. This particle, in addition to being a possible carrier of the fifth force, is characterized by other important properties which make it unique in the world of elementary particles. For instance, as we shall see later, it is the dilaton which controls the precise numerical value of Newton's gravitational constant.

The name of this particle is suggested by its close connection with the "dilatation" symmetry (or conformal symmetry) characterizing the dynamics of elementary but one-dimensionally extended objects (simply called "strings"). Due to this symmetry, the string dynamics should not be influenced by a change of length scale, or energy scale, or time scale, and then by a possible dilatation (or contraction) of the string itself.

The dilaton, which unavoidably appears in all string models as one of the components of the fundamental state of a quantized string, would tend to violate the conformal symmetry producing what is called a "conformal anomaly," that makes the model formally inconsistent. Such a violation can be avoided provided that the dilaton satisfies appropriate differential conditions which—in turn—completely fix its dynamics. It is precisely this set of conditions that determines the role of the dilaton also in the context of the gravitational interaction.

According to such conditions, the dilaton can interact with all existing particles (not only with those carrying baryonic charges, like the graviphoton); such interactions, however, are not universal, namely they may have different intensities for different particles, and the overall intensity depends on the corrections that unavoidably arise when including the effects of quantum physics.

At the present stage of knowledge, unfortunately, our mastery of string theory is not good enough to give us rigorous mathematical methods for an exact computations of the quantum corrections in the regime of very strong interactions: all we can do, at present, is thus to explore the consequences of a few (reasonable) conjectures. There are indications,[14] for instance, that the dilaton coupling constant could be

---

[13] A scalar particle is represented by a mathematical function that has only one component, which is left invariant under a general change of coordinates. A vector particle, instead, is represented by a function with many components, transforming as the components of a vector under a coordinate transformation.

[14] See in particular a paper by Taylor and Veneziano [9].

from forty to fifty times more intense than Newton's constant in the case of heavy particles such as protons and neutrons, and of the same order as Newton's constant for light particles (the so-called leptons), such as muons, electrons, neutrinos.

If this is the case, then the dilaton force acting on a macroscopic body composed of many particles, of various types, tends to be *stronger* than gravity, and dependent on the chemical composition of the body. It is important to stress, also, that the dilaton force is always *attractive*: thus, contrary to what happens with the graviphoton, the overall gravitational interaction turns out to be enhanced, rather than weakened, by the dilaton contribution.

However, even in this case (like in the case of the graviphoton), the absence of any experimental confirmation down to distances of tenths of a millimeter implies that the range of the dilaton force has to be small enough, corresponding to a sufficiently large value of the dilaton mass: in particular, a mass larger than (or at least equal to) about 1000ths of electron volts.

A mass like this would be, nevertheless, very small if compared to the typical masses of other elementary particles: the proton mass, for instance, is a trillion times larger than this tiny mass scale. It is worth mentioning, in this respect, that if the dilaton is lighter than the proton we can then expect new interesting effects occurring at the astrophysical and cosmological level.

For instance, if the dilaton mass is smaller than at least one tenth of the proton mass,[15] then the dilatons produced in the very early, primordial epochs can survive until today in the form of relic cosmic radiation.[16] The detection of such a relic background would give us first-hand information on the state of the early Universe, and on the fundamental interactions in the regime of very high energies.

It would therefore be important to know the value of the dilaton mass. Unfortunately, however, our present understanding of the formal aspects of string theory is inadequate to this purpose: the mass (like the coupling constant) depends, in fact, on the behavior of the quantum corrections in the regime where the dilaton interactions are so strong as to make ineffective the approximate mathematical methods currently at our disposal. The mass of the dilaton thus remains, for the moment, an unsolved mystery.

2.1.2 "Chameleons" and "Fat" Gravitons

As anticipated at the beginning of this chapter, possible anomalies in the gravitational law at small distances could signal not just the presence of new forces but,

---

[15]If the dilaton mass is larger than this limit then the dilaton lifetime becomes smaller than present age of the Universe. All the dilatons copiously produced immediately after the Big Bang thus decay (into photons) before reaching the present epoch, and there is today no relic dilaton background waiting for a possible detection.

[16]The possible existence of a cosmic background of relic dilatons was first suggested and studied by Gasperini and Veneziano (see, e.g., [10]).

rather, new intrinsic properties of the gravitational interaction itself. This second possibility, which at first sight would seem to be more conventional and less innovative than the previous ones, actually leads us to scenarios which are even more "exotic" than those so far considered.

A first example of such a possibility concerns the so-called model of "chameleon gravity."[17]

In this model, one of the contributions to the total gravitational interaction comes from a scalar particle whose effective mass has not a constant value, but a value which depends on the local density of the matter surrounding the particle. This mass, in particular, turns out to be very large near heavy bodies and very small in vacuum, just like a chameleon which adapts to the environment for camouflage! In fact, the closer are the bodies the larger is the mass of this particle, the shorter is the range of the induced force, and the more difficult its experimental detection.

In order to behave in this way the chameleon particle must be characterized by a proper energy which is affected by the interaction with the surrounding bodies. And this is accomplished provided the chameleon couples to the other particles in a complicated "non-minimal" way, through a distortion of the space–time geometry.

Such a coupling looks a bit like the non-universal coupling of the dilaton we mentioned earlier, but with two important differences. The first is that the effective potential energy of the chameleon has a form very different from that of the dilaton; the second is that the coupling properties of the chameleon must be postulated ad hoc, instead of being derived from the very general symmetry principles as in the case of the dilaton.

If this is the case, why should we invent a particle whose main property seems to be just the ability of evading all direct experimental searches? One might say that the theoretical physicists, this time, have made an excessive use of their imagination. It would be so, indeed, were it not for the downside of the problem, namely for the ability of the chameleon to be extremely light in empty space.

In fact, the presence of an extremely light scalar field, carrier of a force with a range large enough to contribute to the gravitational interaction on cosmological scales of distance, is regarded today highly desirable for a possible explanation of some "strange" properties of the present Universe: in particular, its state of cosmic acceleration (that we shall discuss in Chap. 3). On the other hand, all particles characterized by a constant mass and by interactions of gravitational strength should mediate forces of short-range type, not to contradict existing experimental tests: hence, they can hardly affect the cosmic dynamics on large scales. The chameleon, which has a variable mass (and a corresponding variable range) could solve this problem.

The mass of the chameleon, besides varying in space—small on large distances, big on small distances—may also change in time. In fact, the density of cosmic matter decreases in time with the expansion of the Universe, and this unavoidably leads to a reduction of the effective potential energy and of the mass of the

---

[17]See for instance the paper by Khoury and Weltman [11].

chameleon at a cosmic level. The range of the corresponding force will thus increase in time, and when the range will exceed the Hubble radius of our visible Universe, the chameleon will turn into a "normal" cosmological particle, with a mass insensitive to the surrounding matter density.

Another weird gravitational effect, probably even more strange than the chameleon case just considered, is the case of the so-called fat gravitons.[18]

The basic assumption of this model is that the gravitons—i.e., the particles mediating the gravitational interactions—are not point-like objects but are characterized by a finite extension. Their possible "width" is controlled by a very small, but finite, fundamental length L_g.

What are the reasons for this assumption? Mainly the hope of avoiding the formal problems of the gravitational theory associated with the presence of infinitely high values of the energy (the so-called ultraviolet divergences). Such divergences unavoidably appear when extending to the quantum regime a classical theory of gravity like Einstein's theory of general relativity: one finds that such infinities cannot be eliminated, and that the corresponding quantum theory is not consistent.[19]

If gravitons are "slim," or point-like objects, then in any elementary interaction process they can approach other particles at smaller and smaller distances, and exchange higher and higher energies which tend to infinity when the distance tends to zero.[20] If gravitons are "fat," on the contrary, they can approach another particle at most up to distances of the order of L_g, and can thus transmit with their interactions a maximum energy of the order of $E_g = 1/L_g$.

Such a new upper limit—or, using common terminology, such an "ultraviolet cutoff"—of the gravitational energy, due to the graviton "waistline," also controls the quantum contribution to the gravitational energy density of the vacuum. In other words, it produces an effective cosmological constant Λ_g given by $\Lambda_g = E_g/L_g^3 = 1/L_g^4$, completely determined by this new length scale L_g. Such a constant should replace, for all purposes, the one based on the Planck length L_P and given by $\Lambda_P = 1/L_P^4$, which, instead, is to be introduced ad hoc to avoid divergences in models of quantum gravity based on point-like gravitons (see also Sect. 3.3).

Since the Planck length is very small (recall that $L_P \sim 10^{-33}$ cm), the corresponding cosmological constant Λ_P is a huge number: if we call "Planck mass" M_P the inverse of the Planck length we obtain, in fact, $\Lambda_P = M_P^4$, with a value of M_P which is about 10^{19} GeV, namely ten billion billion times the mass of the proton. Such a value of the cosmological constant—as we shall discuss in Chap. 3—is unacceptably large, and in complete disagreement will all observations.

---

[18]See the paper by Sundrum [12].

[19]Usually the infinities appearing in the quantum regime can be eliminated by applying the so-called "renormalization" procedure. In the case of general relativity such a procedure does not work.

[20]In fact, as already mentioned in Chap. 1, the Heisenberg principle tells us that, in a quantum context, the energy exchanges are inversely proportional to the involved distances.

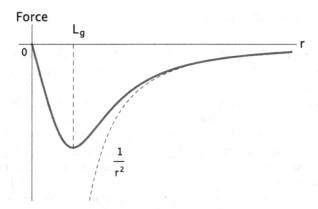

Fig. 2.1 According to the model with fat gravitons, below a critical distance $r = L_g$ the gravitational force deviates from the usual inverse-square law behavior (illustrated by the *dashed curve*) and goes to zero. However, the force stays negative since the interaction is always attractive

The cosmological constant $\Lambda_g = 1/L_g^4$, instead, could be much smaller, provided L_g is much larger than the planck length L_P. The question we should ask is thus the following: which value may (or must) be associated to the length scale L_g?

The answer is given by the experimental tests of the gravitational force at small distance. The model with fat gravitons, in fact, predicts large violations of classical Newton's force as soon as the distance between two bodies fall below the critical value L_g. The effective force, in particular, must go to zero when the distance goes to zero, instead of following the inverse-square law predicted by Newton (see Fig. 2.1).

Such an evident violation of Newton's law should clearly appear in the experiments, unless the length scale L_g is definitely *smaller* than the length scales presently accessible to direct observations (which, as repeatedly stated, are currently limited to about 10^{-2} cm). Since no effect has been observed to date, we must conclude that $L_g \leq 10^{-2}$ cm.

The above conclusion is nonetheless interesting, as it leaves open an exciting possibility. If the value L_g would be just below the present experimental limit, or—more precisely—if it would be about twenty *microns*, equal to 2×10^{-3} cm, then the corresponding cosmological constant would be of the order of $\Lambda_g = 1/L_g^4 \sim (10^{-2}\,\text{eV})^4$, namely exactly of the order of magnitude of the presently observed cosmological constant!

It could be possible, therefore, that the large scale properties of our Universe turn out to be determined by the microscopic properties of the gravitational interactions, and in particular—in the case at hand—by the graviton "waistline." Could that be really so? The model with fat gravitons—like other models—can be confirmed (or disproved) by precise measure of the gravitational force at small distances. Future experiments will thus be able to give us an answer.

2.2   New Dimensions of Space?

Have you ever wondered why, in Newton's law of gravity, one exactly finds "the square" of the distance? Why not another power, different from two, in the exponent of the distance?

It might seem a meaningless question. The answer, however, is instructive.

In Newton's theory the gravitational field satisfies the so-called Poisson equation, which relates the force to the mass density of the gravitational sources. If we consider a portion of space bounded by a spherical surface of arbitrary radius r, and apply a famous mathematical theorem, the Gauss theorem, then the Poisson equation tells us that the gravitational force, multiplied by the area of the spherical surface, $2\pi r^2$, turns out to be proportional to the total mass contained inside the sphere. Dividing such a result by the surface area then one immediately finds that the gravitational force is proportional to the mass, and inversely proportional to square of the distance.

The Poisson equation and the Gauss theorem are valid in spaces with any given number of dimensions. The above reasoning could thus be repeated in an identical manner imagining to live in a world where the number of spatial dimensions is larger than three. Suppose, for instance, that there are N dimensions: the result of our argument would be similar to the previous one, but with one important difference.

In order to bound a given portion of a three-dimensional space we need to use a closed two-dimensional surface (for instance a sphere). In order to bound a portion of a N-dimensional space we need instead a closed hypersurface with dimensions not 2 but $N-1$! (for instance, not a sphere but what is called a "hypersphere"). The area of a sphere is proportional to the square of its radius, r^2, the "area" of a $(N-1)$-dimensional hypersphere is proportional, instead, to r^{N-1}. Dividing by the generalized area of this hypersurface we then find that the gravitational force turns out to be proportional to the mass (as before), and inversely proportional not to the square of the distance, but to the distance raised to the power $N-1$.

Hence, the fact that the gravitational force turns out to depend on the inverse square of the distance is closely connected to the fact that the world in which we live has just three spatial dimensions. The same conclusion is also valid for the case of Coulomb's law, describing the attraction (or repulsion) of static electric charges, quite similarly to the case of Newton's force.

Such a result on the number of spatial dimensions ($N = 3$) is hardly surprising, on one hand (as we are well accustomed to three dimensions from everyday experience); on the other hand, it complicates our work of theoretical physicists. In fact, there are two ingredients which may greatly simplify the formulation of a unified model including all fundamental interactions: one is supersymmetry (already mentioned in the previous sections), the other is a space with more than three dimensions.

The idea of using additional spatial dimensions as crucial ingredients of a unified theory was born nearly a century ago, and traces back to the works of Kaluza and Klein [13]. They have shown in their papers that the Einstein theory of general relativity, if formulated in the presence of one additional spatial dimension, may be interpreted—under appropriate conditions—as a theoretical model including not only the gravitational interactions but also the electromagnetic ones, both described in ordinary three-dimensional space.

This idea (suitably generalized) has later received a definitive consecration with the rise of string theory, in the 1980. In string theory, in fact, the additional dimensions are not optional ingredients, as they are essential (as we shall see later) to obtain models proving to be physically and mathematically consistent.

If we want to take seriously the theoretical schemes with more than three spatial dimensions, however, we must answer a simple—yet crucial—question: why the space in which we live appears to be three-dimensional? what happened to the other dimensions (also called "extra" dimensions), if they really exist?

There are two possible types of answer to these questions.

The first answer is that, until now, we have experienced only three dimensions because the additional dimensions are extremely small, and "rolled up" upon themselves (or, following the usual mathematical terminology, they are "compactified" on very small length scales). To see them "rolled out" we should perform experiments at energy scales so high to be out of reach for current technology.[21]

The second, possible answer is that we are not aware of the additional dimensions not because they are confined in very small regions of space, but simply because the fundamental forces we are using to probe the space in which we are immersed, and to interact with the world around us, propagate only around three dimensions. The extra dimensions do exist, but they elude our direct sensorial (and experimental) experience, just like a physical phenomenon occurring outside the limited sensitivity band of our senses (or of our technological instruments).

The presence of additional spatial dimensions, whether they are compact or impenetrable, can however modify the behavior of the gravitational force at small distances, as we shall see in the following sections. Such modifications could bring us information, indirectly, also about those dimensions otherwise inaccessible to a direct observation.

2.3 The "Compact" Scenario

Let us start with the case in which we are living in a space with extra dimensions of very small and compact "shape." We will see below how such a configuration can be arranged, thanks to a peculiar mechanism called "spontaneous compactification."

---

[21]The search for new additional spatial dimensions is one of the main goals of the big accelerator LHC at the CERN laboratories, near Geneva. At the time of writing (March 2013) no positive result has yet been reported. A final absence of signals would imply that the additional dimensions, if present, may become visible only at energies higher than the maximum values accessible to LHC (i.e., about 14 TeV).

Fig. 2.2 An example of Kaluza–Klein geometry for a two-dimensional space with one compact-ified dimension. A very long cylinder with tiny diameter, if observed from a sufficiently large distance, may appear as a one-dimensional object, extended in length but with no thickness

In order to visualize a geometric configuration of this type we can imagine a very long and thin cylindrical object (see Fig. 2.2). Its surface corresponds to a two-dimensional space, and one of its dimensions—the one aligned along the cylinder axis—can be extended in length without limits; the other dimension, instead, is curled up onto itself in a compact shape (a circle of finite radius). If the radius is small enough, and we are looking at the cylinder from a sufficiently large distance, the cylinder will look like a one-dimensional object, in every respect. Consider, for instance, a hair: to the naked eye it appears to be extended only in length, and we need a microscope to appreciate its transverse thickness.

Our Universe, according to the Kaluza–Klein model, could be characterized by a similar geometrical structure. Namely, by three spatial dimensions indefinitely extended (like the length of the above-considered cylinder), plus other dimensions with a small and compact shape. Along such extra dimensions it is only possible to move in a circular path, always coming back to the starting point after traveling very short distances. If we are not equipped with powerful enough instruments, able to resolve (directly or indirectly) distances of the order of the radius of the compact dimensions, then we will only observe the three dimensions extended on the macroscopic scales.

As we have anticipated, the role of the extra dimensions, in this context, is that of allowing a geometric representation of the fundamental interactions, with the final purpose of including all of them, together with gravity, in a complete unified theory. To that purpose, which other property (in addition to a compact shape and a very small size) has to be met by the extra dimensions?

Let us consider the simplest case in which our space has only one additional dimension, and let us call it the "fifth dimension" (as we take into account also the presence of the temporal dimension, besides the three spatial ones). According to the original Kaluza–Klein idea, the gravitational field acting along the fifth dimension should play the role of the electromagnetic field in ordinary three-dimensional space. This is possible provided the five-dimensional model satisfies an important geometric constraint: the symmetry properties of the electromagnetic field—among which, in particular, the so-called gauge invariance[22]—must correspond to suitable

---

[22]The gauge symmetry is a property ensuring that the electric and magnetic fields remain unchanged under suitable transformations of the electromagnetic potential.

symmetry properties—called "isometries"—of the five-dimensional space–time manifold that we are considering.

In the approach adopted by Kaluza and Klein such a requirement is satisfied by assuming for the space a factorized geometrical structure, i.e., by assuming that the whole space can be represented as a product of two subspaces: one is the usual (infinitely extended) three-dimensional space, and the other is a compact one-dimensional space, of (very small) radius L_c. In such a context, all physical variables included in the model—hence, in particular, those describing the gravitational field—can be factorized as the product of two functions: one depending only on the three coordinates of ordinary space, and the other one depending only on the fifth coordinate (and being necessarily a periodic function, as the fifth dimension is similar to a circle).

Within such a geometrical structure, any model describing interactions of pure gravitational type (for instance, the theory of general relativity), if formulated in five dimensions, automatically separates into two parts: one representing the gravitational interactions in ordinary three-dimensional space, and the other one representing—in the same space—the electromagnetic interactions among charges. Hence, the model successfully achieve the goal of describing the two interactions in a geometric and unified way.

What is perhaps most interesting in this—well thought—unified scheme, however, is the fact that the electromagnetic and gravitational forces predicted by this model are not exactly the same as those separately predicted by the usual (not unified) models of these two interactions. The differences, mainly arising in the small-distance and/or high-energy regimes, are of two types.

2.3.1 The Kaluza–Klein "Towers" and the Radion

A first difference is due to the fact that all particles appearing in a Kaluza–Klein model—hence, in particular, the graviton and the photon—are associated to an infinite series of "partners," represented by massive, very heavy particles. For any given particle the set of all partners, whose mass spectrum is growing with discrete and equally spaced steps, is called a Kaluza–Klein "tower."

The mass of these new particles is due to the presence of the compact dimension, grows proportionally to the positive integer numbers $(1, 2, 3, \ldots)$, and is inversely proportional to the compactification radius L_c (see Fig. 2.3). Given that L_c is very small, these masses are very large, hence the range of the forces associated to these new particles is very small. In practice, the corrections induced on ordinary electromagnetic and gravitational forces become important only for distances of the order of L_c.

The question that obviously arises, at this point, is the following: how large (or small) is L_c?

We cannot give a precise answer to this question, unfortunately, because a generic Kaluza–Klein model allows in principle arbitrary values for this parameter. We

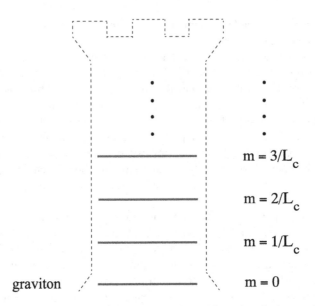

$$m = 3/L_c$$

$$m = 2/L_c$$

$$m = 1/L_c$$

graviton $m = 0$

Fig. 2.3 An example of Kaluza–Klein tower: at the ground floor there is the massless graviton, on the upper floors the infinite series of its massive partners, with masses depending on the compactification radius L_c

know however that L_c, by definition, controls the ratio between the strength of the gravitational force in the four-dimensional space of the Kaluza–Klein model and the corresponding strength in ordinary three-dimensional space. If these strengths are the same, then L_c has to be of the order of the Planck length: $L_c \simeq L_P \sim 10^{-33}$ cm. If, on the contrary, the gravitational force in the higher-dimensional space is stronger than in three dimensions, then L_c must be larger than L_P.

In any case, in order to be acceptable, the value of L_c must be small enough to have not produced any detectable effect in all the experiments so far performed. This applies both to the direct measures of the gravitational force (currently exploring distances down to the scale of 10^{-2} cm) and to the experiments studying high-energy particle collisions (that in the LHC accelerator, at CERN, are sensitive to distances of the order of 10^{-16} cm).

A second, interesting difference between the unified Kaluza–Klein model and the usual theory of electromagnetic and gravitational interactions, arises from a possible distortion (in time and space) of the geometry of the fifth dimension. This effect, typically induced by the close connection existing between gravitation and geometry, leads to fluctuations of the effective compactification radius, and such fluctuations—if not suitably stabilized—are represented by a scalar particle called the "radion."

The Kaluza–Klein scenario thus introduces in general a new particle, the radion, with physical properties very similar to those of the dilaton already met in the previous sections, but with a different geometric origin. Like the dilaton, the radion is directly coupled to the electromagnetic field, and should thus produce an effective gravitational force which is different for bodies with different internal electromagnetic structure (namely, with different chemical composition).

The resulting effect is a strong violation of the equivalence principle which should be observable if radions were massless particles, as predicted by the simple version of the Kaluza–Klein model. Hence, such a model must be constrained in order to stabilize the fluctuations of the compactification radius (thus eliminating the radion); alternatively, the model is to be generalized so as to give a mass to the radion, and make that particle sufficiently heavy.

2.3.2 A "Spontaneous" Compactification

Adding only one extra dimension was enough at the times of Kaluza and Klein, when the only interaction to be "geometrized" (i.e., to be included into a unified scheme together with gravity) was the electromagnetic interaction. Today, however, this would not be sufficient any longer, as we are aware of the existence of other fundamental interactions (the weak and strong nuclear forces), and their inclusion would require the addition of other extra dimensions.

We might think of a natural generalization of the original Kaluza–Klein model obtained by replacing the "circle" of the fifth dimension with a compact higher-dimensional space. By doing so, however, we come across a problem.

In fact, as we have already stressed, the geometric properties of the extra dimensions must reflect the symmetry properties of the associated interactions. In the five-dimensional case discussed so far, in particular, we have used a geometry compatible with the Abelian gauge invariance of the electromagnetic field. The new interactions we want to include, however, are characterized by a more complicated, non-Abelian type of gauge invariance. Hence, the geometry of the associated higher-dimensional space should be characterized by non-Abelian symmetry properties.[23]

Now, it is certainly possible to add to the ordinary three-dimensional space a new, very small and compact space which has an arbitrary number of dimensions and a geometry compatible with isometries—namely, with symmetry transformations—of non-Abelian type. In that case, however, also the three-dimensional geometry becomes highly curved and distorted! It is impossible, in that case, to reproduce "realistic" geometric configurations resembling the world we live in, namely a word where three spatial dimensions are nearly flat and macroscopically extended.

The spatial configuration we would like to achieve, with three flat dimensions and all the other ones curled up in a space of very small and compact size, is only compatible with a special type of geometry satisfying the condition of zero "Ricci curvature." This condition, in turn, is only compatible with symmetries of

---

[23]Two (or more) symmetry transformations are called Abelian if they give the same result regardless of the order in which they are performed. Otherwise they are called "non-Abelian."

Abelian type. This implies that the strong and weak nuclear interactions (of non-Abelian type) cannot be geometrically represented as gravitational effects in some appropriate extra-dimensional space, but are to be added to the model in a non-geometric way.

This conclusion, on one hand, ruins the simplicity and the elegance of the original Kaluza–Klein idea, namely the possibility of describing all interactions through a theory of pure gravity formulated in a higher-dimensional space–time. On the other hand, the addition of "extra" (i.e., non-gravitational) fields, possibly associated to non-Abelian interactions, provides a mechanism to explain "why" the extra dimensions are not spatially extended like the three dimensions of our macroscopic space: the so-called mechanism of "spontaneous compactification."

Such a mechanism is triggered indeed by the presence of additional, non-geometric fields, characterized by the same factorized structure as that of the higher-dimensional geometry we want to achieve. More explicitly, they are characterized by an energy density—or better, an "energy-momentum tensor"—which can be splitted into two distinct parts: one depending only on the three dimensions of ordinary space, and the other one depending only on the extra dimensions of the additional compact space (quite independently of their number). No cross-terms are present.

This type of energy-momentum distribution can be achieved by the fields describing the non-Abelian interactions, but also by other objects like antisymmetric tensors, monopoles, and so on. The space–time geometry, on the other hand, is rigidly constrained by Einstein's equations which imply, in any number of dimensions, that the geometric properties have to be tuned on the properties of the matter sources present in the model. It is thus the presence of the appropriate matter fields that prescribes to some dimensions to be compact (or not), and that fixes the possible radius of compactification. "Spontaneously" or not, all spatial dimensions have to comply with this mechanism.

Let us conclude this section with an important remark concerning the size of the volume of the compact extra-dimensional space.

In a model of Kaluza–Klein type, and in the presence of a compact n-dimensional space of radius L_c, the ratio between the strength of the gravitational force in the higher-dimensional space and in the ordinary three-dimensional space is controlled by the size of the compactification volume, namely by L_c^n. Consider, for instance, a Kaluza–Klein model formulated in space–time with a total number $D = 4 + n$ of space–time dimensions, and call G_D the gravitational constant of this higher-dimensional space: one then finds that G_D is given by the usual Newton constant G multiplied by the compactification volume L_c^n.

The Newton constant G, on the other hand, can be expressed in terms of the Planck length squared as $G = L_P^2$ (by definition of L_P). In a similar way, we can introduce a new length scale L_D, typical of the higher-dimensional space–time, such that $G_D = L_D^{2+n}$. This expression holds together with the previous one relating G_D to the compactification radius, namely $G_D = L_P^2 L_c^n$.

Suppose now that the gravitational strength is nearly the same in any number of dimensions, namely, that the two length scales L_P and L_D coincide, or are at least of the same order: it follows that the compactification volume must be characterized by a mean radius which is also of the order of the Planck length, $L_c \simeq L_P$ (and this is an extremely small size, totally out of reach for current technology). However, if the gravitational strength in D dimensions is larger than in four dimensions, then also the extra-dimensional volume, and the corresponding compactification radius L_c, could have a size larger than Planckian.

Reversing the argument, we can exploit the present experimental limits on the size of the compact dimensions to get information on the allowed number n of such dimensions, and on the gravitational coupling constant of the higher-dimensional space.[24]

We know, in particular, that the mean compactification radius L_c should be smaller than about 10^{-2} cm, to avoid inconsistencies with the direct measures of the gravitational force so far performed. Is it possible, without violating this bound, that the strength of higher-dimensional gravity is so large as to produce effects detectable at least by the most powerful accelerators?

The most powerful experimental apparatus is currently available at the CERN laboratories, and is the LHC accelerator which—as already mentioned—can explore the high-energy domain up to the TeV scale. Thus, to be within the reach of the LHC sensitivity, the coupling strength G_D of higher-dimensional gravity should be characterized by an energy scale $1/L_D$ which of the order of one TeV (instead of being of the order of the Planck energy scale, typical of the gravitational strength in three-dimensional space).

Once L_D is fixed to the TeV scale, if we use the relations connecting the coupling constants G, G_D and the compactification volume, and impose on the compactification radius the experimental constraint $L_c \leq 10^{-2}$ cm, we then find that higher-dimensional gravity could be strong enough to produce detectable effects in the LHC experiments provided there are at least $n = 2$ (or more) extra dimensions.

We should also take into account that, up to now, there are no hints of extra dimensions not only in the gravitational experiments, but also in all experiments of high-energy particle collisions, probing small distances down to the scale of 10^{-15} cm. Imposing this second, more stringent constraint on L_c, it follows that the possibility of the TeV scale as a typical energy scale of higher-dimensional gravity is still open, provided that the number of extra dimensions is sufficiently high: larger (or at least equal) to $n = 15$. If n is smaller, the energy scale $1/L_D$ typical of higher-dimensional gravity has to be higher than the TeV scale and thus, unfortunately, beyond the reach of LHC.

However, there is a possibility to evade such (rather negative) conclusions if we rely on a different type of higher-dimensional scenario: the so-called brane-world scenario, which will be illustrated in the following section.

---

[24]See for instance the paper by Arkani Hamed et al. [14], and the paper by Antoniadis [15].

2.4 The "Brane-World" Scenario

A second possibility to make experimentally "invisible" the extra dimensions, avoiding their confinement in regions of space of very small and compact size, relies on the idea that the fundamental interactions propagate uniquely—or at least primarily—along the *three* spatial directions.

In fact, the tools we are using to explore the space around us—from our eyes up to the most powerful and sophisticated technological instruments—are all working on the basis of the fundamental interactions. If such interactions were to exploit only a reduced number of dimensions among all the available ones (just like waves propagating on the surface of a stretch of water, and never in the direction perpendicular to it), then the other dimensions would remain hidden to all practical purposes.

Our physical experience could be confined, in this way, to a three-dimensional "slice" of the whole space. Such a slice—called "three-brane," which means three-dimensional membrane—would represent, in practice, the only portion of space accessible to our direct exploration.

The idea is attractive, but—like all other ideas used in physics—can be seriously considered in a scientific context only if appropriately motivated, and formulated in the context of a complete and quantitative theoretical scheme. In our case, a model of the three-dimensional world as a brane embedded in an external higher-dimensional space—also called "bulk" space—is indeed suggested (and made possible) by string theory.

In order to illustrate this point let us recall that a string is an elementary (but not a point-like) object, with an intrinsic (finite) extension along one spatial dimension (we can think of a string as a very small and infinitely thin twine). To describe the motion of a string in a mathematically complete and consistent way we must specify not only its initial position, but also the so-called boundary conditions carrying information on the position and the possible velocity of its endpoints.

There are two possible types of boundary conditions. The Neumann conditions, allowing the endpoints to move (in such a way, however, that the kinetic energy does not leave the string flowing outwards through the endpoints). And the Dirichlet conditions, imposing on the endpoints to stay fixed.

The boundary conditions are to be specified along all the dimensions of the space in which the string is embedded, but they are not necessarily required to be all of the same type. In a three-dimensional space, for instance, we could impose the Neumann conditions along the two dimensions spanned by the cartesian coordinates x and y, and the Dirichlet conditions (i.e., fixed endpoints) along the third dimension, corresponding to the z coordinate.

In that case we would obtain a configuration where the string endpoints may assume arbitrary positions along the x and y directions, but are always fixed at the same position ($z = $ constant) along the z direction. The result (illustrated in Fig. 2.4) is that the string endpoints stay confined on a spatial slice called "Dirichlet brane" (or, more shortly, "D-brane") which, in the case we are considering, is a

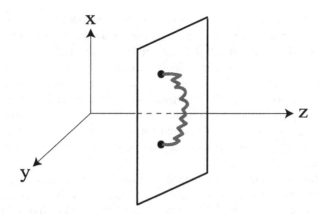

Fig. 2.4 An example of two-dimensional Dirichlet brane. The string endpoints can freely move on the brane (which, in this case, coincides with the $\{x, y\}$ plane shown in the figure), but they are always at the same fixed value of z

simple two-dimensional surface corresponding to the Euclidean plane $\{x, y\}$ located at $z = $ constant.

As we shall see later in more detail, the space in which a string is embedded must be characterized by more than three dimensions, in order that the string model may prove to be physically and mathematically consistent. In such a higher-dimensional context we can thus impose the Neumann conditions (moving endpoints) along three spatial dimensions, and the Dirichlet conditions (fixed endpoints) along all other existing dimensions. The result will be similar to the previous one, with the only difference that the string endpoints will now be localized on a three-dimensional Dirichlet brane (three-brane), which can be interpreted as the space corresponding to our usual three-dimensional world.

We are now in the position of explaining why (and how) string theory allows fundamental interactions to be confined in three spatial dimensions.

In the string models providing a unified description of all interactions, in fact, the elementary charges sourcing the (Abelian or non-Abelian) gauge fields are localized just on the endpoints of the open strings. If such endpoints are confined on a Dirichlet three-brane, then the charges are also confined, and we can then obtain a model where the corresponding (Abelian or non-Abelian) gauge interactions propagate only along the three-dimensional space of the brane.

There is a particular interaction, however, that can evade this rule: the gravitational interaction. Why just gravity?

The reason is simple: the source of gravity is the total energy (which is not a charge) so that, in the context of string theory, gravity is associated not to open strings but to closed strings (very small twines forming a loop, like an infinitely thin rubber band). Closed strings automatically satisfy the boundary conditions, and there are no free endpoints to which apply the Neumann or Dirichlet conditions, hence they can freely propagate through the whole higher-dimensional space.

In view of this aspect of string theory, it might seem that the model of a three-brane embedded in a higher-dimensional space should be immediately rejected, being in marked contradiction with experimental observations.

As already stressed, in fact, in a higher-dimensional space the gravitational force does not follow the inverse-square law: it decreases with the distance like the distance raised to a power different from two (in general, a power equal to the total number of spatial dimensions minus one, see Sect. 2.2). So, if we want to save the "brane-world" scenario, we should find a way to evade this result which would contradict the experimental measures of the gravitational forces on macroscopic scales of distance.

A first (obvious) solution is provided by the assumption that all dimensions external to the three-dimensional brane are characterized by a very small and compact size, exactly as in the case of the extra dimensions of the Kaluza–Klein models previously discussed.

There is, however, another interesting solution, based on the fact that the special geometry of the brane-world scenario seems to be able to confine the forces on the brane even in the case of the gravitational interaction. Such an effective confinement, as we shall see, is only partial, but is enough to avoid phenomenological inconsistencies, and to make the model fully compatible with the presence of extra dimensions even if they are arbitrarily extended.

2.4.1 The Geometric Confinement of Gravity

We should recall, in fact, that a three-brane (like any brane and any other physical object) contains an intrinsic energy density which tends to bend the geometry of the surrounding space, according to the laws of Einstein's theory of general relativity. The space of the brane may remain flat, provided that the external space acquires a curvature described—at least in the simplest case of the Randall–Sundrum model[25]—by the so-called anti-de Sitter geometry.

As a consequence of such a geometric structure, the gravitational force produced by the masses localized inside the brane separates into various components. One component is of long-range type and is mediated by the usual massless gravitons; in addition, there is an infinite number of short-range components, mediate by massive particles, whose mass grows (in a continuous way) from a minimum, finite value up to infinity.

Such a situation closely resembles the towers of Kaluza–Klein particles: indeed, we have again an infinite series of massive partners of the graviton, and a resulting modification of the effective gravitational force. There are, however, various important differences.

---

[25]Presented in a famous paper by Randall and Sundrum [16].

A first (and probably most important) difference is that, in the brane-world model, only the massive particles can propagate along the extra dimensions. The massless component of the gravitational force is "trapped" into the brane, prisoner of a bottomless well (corresponding, in practice, to an infinitely deep and narrow potential well): a massless graviton cannot escape from that well, not even by resorting to quantum effects. The long-range gravitational force keeps rigidly confined on the three-dimensional space of the brane, and thus follows exactly the expected Newtonian (i.e., inverse-square law) behavior.

Another important difference concerns the short-range forces (that can evade the mechanism of geometric confinement, and that propagate through all spatial dimensions). The strength of these forces, mediated by the massive partners of the graviton, is not universal but depends on the mass of the carrier particle: at low enough energy, in particular, the coupling strength grows proportionally to the mass of the particle. The mass spectrum of the graviton partners, in its turn, varies in a continuous way,[26] thus producing a continuous variation of the effective coupling parameter.

Nevertheless, it is possible to estimate the overall contribution of all these new components of the gravitational interaction (at least in the approximation in which the interaction is sufficiently weak), and to compute the resulting correction to the ordinary Newton's law.

In the simplest case in which there is only one additional dimension external to the brane, for instance, one finds that the overall effect produced by the massive gravitons is described by a new force varying as the inverse of the fourth power of the distance. The intensity of such a new force becomes comparable with the intensity of the usual Newtonian component at distances of the order of the curvature radius of the space external to the brane; at larger distances, the new force becomes rapidly negligible.

This result is interesting (and different from previous results obtained in the context of the Kaluza–Klein models) because the corrections to the Newton force are controlled by the *curvature radius* (and not by the compactification radius) of the extra dimensions. Namely, they are controlled by the local geometric properties of the extra dimensions, not by their global shape or extension. It follows, in particular, that if the extra dimensions are sufficiently curved (in particular, if their curvature radius is not larger than about 10^{-2} cm), then they are allowed to be even arbitrarily extended without contradicting the existing experimental evidence.

Could it be really so? Namely, are we living in a three-dimensional world embedded, like a brane, into an external (bulk) space with more than three dimensions?

According to the brane-world scenario, suggested by string theory and briefly presented in this section, the future measurements of the gravitational force—increasingly accurate, probing smaller and smaller distances—might be able to give us convincing answers.

---

[26]In the case of the Kaluza–Klein models, instead, the mass spectrum is discrete (see Sect. 2.3.1).

Chapter 3
Gravity at Large Distances

As discussed in the previous chapter, at small enough distances we may expect several possible modifications of classical Newton's law describing the behavior of the gravitational force. These modifications can be attributed to the entry in a microscopic regime characterized by new symmetries and governed by new (quantum) rules, or to the presence of extra dimensions inducing additional short-range contributions to the gravitational force. None of such modifications, however, has been experimentally confirmed so far.

In the opposite limit of large distances the effects quoted above should disappear, both because the short-range forces become negligible and also because, at large distances, we expect to stay within the scope of the classical gravitational theory (well described by Newton's law, and by its relativistic completion represented by the Einstein equations). It is tempting to say, therefore, that there should be no surprises. At large enough distances, however, several new and striking gravitational effects have already been discovered! and we are still searching for their conclusive and fully satisfactory explanation.

I am referring, in particular, to two well-known effects occurring on astrophysical and cosmological scales. The first one, discovered more than 40 years ago,[1] concerns an anomaly in the gravitational field of (almost) all galaxies: the field is stronger than it should be, and forces the stars to orbit around the galactic center faster than expected.

Let us recall, in fact, that the rotation velocity of a star is associated to a centrifugal force able to counteract the centripetal attraction due to gravity, so as to stabilize the star on its orbit (just as it happens for the planets around the sun). As we move away from the galactic center the gravitational attraction should become weaker and weaker, and the rotation velocity should correspondingly decrease. What is being observed, instead, is that the stars remain rotating at a

---

[1] Mainly thanks to the work of Vera Rubin, a young astronomer of the Carnegie Institution (Washington), who first was able to measure accurately enough the rotation velocity of the stars as a function of their distance from the center, in the cases of many spiral galaxies.

M. Gasperini, *Gravity, Strings and Particles*, DOI 10.1007/978-3-319-00599-7_3,

practically constant rate, just as if the intensity of the galactic gravitational field were independent of the distance from the center.

The second effect, discovered at the end of the last century,[2] concerns an anomalous repulsive force observed on very large scales of distance. Under such a force the expansion of our Universe turns out to be accelerated, instead of being decelerated as expected in the case of a gravitational field generated by the matter sources we know to be present on cosmic scales (stars, galaxies, intergalactic radiation, and so on), and according to the laws of Einstein's theory of general relativity.

It is possible, in principle, that the anomalous effects we have mentioned—the galactic rotation curves and the cosmic acceleration—are signals pointing at the need for changes in the laws of gravity at very large distances. Actually, it is possible to formulate generalized models which are consistent with the results of known gravitational theories at the ordinary scales of distance, and that, at large enough distances, are able to reproduce the observed anomalous effects.

We may quote, in this respect, the model of "chameleon gravity" (already presented in Sect. 2.1.2); the so-called MOND model[3] (able to explain the behavior of the galactic rotation curves through an *ad hoc* modification of Newton's law); the models called "$f(R)$ gravity" (proposing *ad hoc* modifications of the laws of Einstein's gravitational theory[4]); and the models of "induced gravity" (based on the presence of extra dimensions), that will be illustrated in the following section.

However, the most conventional—and most widely accepted, up to now at least—attitude to explain the above observed anomalies is not that of changing the known laws of gravity. The dominant attitude is to assume the existence of two new types of matter and energy present at the cosmic level, which we have not yet been able to produce (or to detect) with experiments performed in our laboratories on Earth, but which have the "right" properties to be responsible for the observed cosmic anomalies.

These new substances have been named, respectively, "dark matter" and "dark energy," as they both would be invisible, were it not for the gravitational effects they produce.

Dark matter, in particular, accumulates around each galaxy like a halo of very tiny cosmic "dust," filling the empty space among the stars and thus producing an increase of their rotation velocity. Dark energy, instead, is distributed like a nearly homogeneous and isotropic fluid throughout the visible Universe, and has

---

[2]The discovery was reported in the papers by Riess et al. and Perlmutter et al. [17]. The work of both these groups has been awarded the Nobel prize in Physics in 2011.

[3]The abbreviation stands for *Modified Newtonian Dynamics*, a model of classic mechanics (proposed by Milgrom in 1981) which deviates from standard Newtonian mechanics only in the limit of very small accelerations. In that limit, a force is no longer proportional to the acceleration, but tends to be proportional to *the square* of the acceleration.

[4]The name of these models is due to the fact that they modify the equations of general relativity by replacing the space–time curvature—represented by the symbol R—with an arbitrary function of the curvature, denoted indeed by $f(R)$.

the strange property of being characterized by a negative pressure: in this way it can be the source of a repulsive gravitational force, strong enough to accelerate the expansion of the whole Universe.

It is granted that a direct detection of dark matter and/or dark energy would constitute a discovery of paramount importance for modern physics. So far there is no clear positive result in this context. I should say, however, that just at the time of writing these notes (today, 3 April 2013), we got a news that could represent an indirect confirmation of the existence of the dark matter.

The experiment AMS has announced,[5] in a public seminar held this afternoon at CERN in Geneva, that the flux of cosmic rays hitting the Earth (coming from our entire galaxy) contains a fraction of antimatter higher than expected (in particular positrons, i.e., antielectrons). Such an excess of positrons is constant in time, and independent of the arrival direction (namely, it is isotropic), in agreement with the results of the previous experiments. Thanks to its very high precision, however, the AMS experiment is also able to conclude that the distribution of the positrons as a function of their energy has exactly the behavior one would expect if the positrons were produced by dark matter particles, colliding and disintegrating among each other!

So, have we obtained a signal confirming—albeit indirectly—the existence of dark matter?

It is too early to provide a positive (or negative) answer. We will need to measure the number of positrons also at higher energies (in particular, higher than the present limit of about 250 GeV), to see whether the surplus of positrons is reduced, and at some point disappears. In fact, if the excess positrons are produced by the disintegration of dark matter particles, then the effect should disappear when considering positrons of energies higher than the rest mass of those particles.

No reduction of the excess positrons has been observed, up to the energy scales so far analyzed. It will be necessary to collect and analyze data for several additional months, before obtaining accurate results at higher energies and—maybe—conclusive answers about the dark matter origin of these positrons.

Leaving aside for the moment dark matter, in the rest of this chapter we will focus on the problem of the cosmic acceleration, with a short discussion of three interesting possibilities. We will consider, in particular, the case of cosmic acceleration due to (1) a modification of the gravitational interaction at large distances; or to (2) the presence of a new substance, or field of forces, with the typical properties of dark energy; or to (3) unconventional physical properties of the state that we call the "vacuum," and that still we do not understand deeply enough.

---

[5]The acronym AMS means *Alpha Magnetic Spectrometer*, an instrument used for the detection and study of the particles of antimatter present in the cosmic rays reaching the Earth. Such an instrument is installed in the outside of the International Space Station (ISS) orbiting around the Earth, and has the task of intercepting the cosmic rays before they can interact with the terrestrial atmosphere (and lose precious information about their origin).

3.1 Extra Dimensions Strike Back

There is a number of models predicting deviations from the standard gravitational dynamics at very large distances, and able, in principle, to explain the present accelerated expansion of our Universe even without resorting to exotic ingredients like dark energy.

Almost all these models, however, are formulated in a somewhat contrived way: they are invented to the only purpose of solving the problem of the cosmic acceleration and, in general, they lack in convincing theoretical justifications. In addition, some of these models are in trouble in describing gravity in the regime of ordinary macroscopic distances (as they predict additional long-range forces which, if not suitably eliminated, are inconsistent with the observational results).

An exception (at least to the first points of the above criticism) is represented by the so-called DGP model, or model of "induced gravity,"[6] based on the brane-world scenario introduced in Sect. 2.4. Indeed, the idea that our space is a three-dimensional slice of a higher-dimensional space is not introduced *ad hoc* to explain the cosmic acceleration: as we have seen, it is suggested by string theory and motivated by a mechanism of spatial confinement of the fundamental interactions.

It might seem strange—and it is actually so—that in the context of a brane model one can manage to obtain deviations from the standard gravitational dynamics at large distances. In the previous chapter, in fact, we have seen exactly the opposite effect: namely, we have seen that the extra dimensions external to the brane, after confining on the brane the long-range component of the gravitational force, may induce corrections to that force only at sufficiently *small* distances.

In that case, however, we were considering a particular geometric configuration characterized by an intense vacuum energy (or cosmological constant) present both on the brane and on the external "bulk" space, and associated to highly curved— even if large—extra dimensions. In that context, the corrections to the gravitational field, on the brane, were possibly important only at very high energies, hence at small enough distances.

In the context of the DGP model, instead, one assumes that all existing forms of matter and energy are strictly localized on the brane, together with all interactions of non-gravitational type. The bulk space external to the brane is empty and with no intrinsic energy, its geometry is flat (of Euclidean type), and gravity is completely free to propagate even along these external dimensions. However, the brane itself has its intrinsic gravitational field (the so-called induced gravity, associated to the matter sources localized on the brane), which contributes to the total gravitational field present in the higher-dimensional space.

The result we obtain, in this way, is just the opposite of the previous one. In such a context, the gravitational field acting inside the brane "becomes aware" of the

---

[6]The abbreviation DGP comes from the names of the authors, as the model was jointly proposed by Dvali, Gabadadze, and Porrati [18].

presence of the extra dimensions (and modifies accordingly) only at sufficiently *low* energies, hence—on a cosmic scale of distances—only at sufficiently recent times. This result is interesting for two reasons.

First of all it shows, with an explicit example, that the presence of a higher-dimensional space may be compatible with the usual behavior of the three-dimensional gravitational interactions, at least in an appropriate energy range, even if the extra dimensions are not only *infinitely extended* but also *flat*. There is no need neither for their compactification (required by the models of Kaluza–Klein type), nor for their curvature (required instead by the models of Randall–Sundrum type).

Second—and this is probably the more stunning and interesting aspect of the DGP model—the predictions for the modified gravitational dynamics on the brane imply that, below an appropriate energy scale, the cosmological expansion of the three-dimensional brane space automatically becomes accelerated! One thus obtains just the desired effect, and finds that it occurs just at the low-energy epochs typical of the current state of our Universe.

Is this the right explanation of the cosmic repulsive force that we are presently observing on large scales? Are we entitled to interpret such an effective force as a hint of the presence of extra dimensions which surround our three-dimensional world, which are infinitely extended and yet so elusive to a direct observation?

I would like to say "yes," but, actually, we have no sufficient evidence yet to give an answer (neither affirmative nor negative). We should recall, also, that the DGP model is not free from problems, of both phenomenological type and formal type. On the phenomenological side, in fact, it requires an unusually large value of the coupling constant which controls the gravitational strength in the higher-dimensional space: such a value is difficult to be justified with the physical arguments. On the more formal side, it seems to be incompatible with a consistent formulation in a quantum-mechanical context.[7]

3.2 A New Form of "Dark" Energy?

Let us suppose that gravity is always well described—without surprises—by Newton's laws in the weak field regime, and by Einstein's laws in the relativistic regime. Suppose that such laws can be applied, without modifications, even at a cosmological level to describe the global evolution of the whole space–time. In that case, why there should be difficulties to justify a phase of accelerated expansion like the one currently experienced by our Universe?

It is probably appropriate to recall, in this regard, how we come to determine the state of cosmic acceleration.

---

[7]Because of the presence of the so-called ghost states, unphysical states are characterized by a negative probability, appearing in the quantum version of the DGP model.

We consider a class of "sample" light sources, i.e., sources emitting light at a known and fixed rate, and then we measure the intensity of the energy flux received by these sources as a function of their distance from Earth. A typical example of sources appropriate to this purpose is represented by Supernovae of type Ia, i.e., by exploding stars emitting a huge amount of light energy, visible at very large distances. The energy is radiated away with an emitting power which is practically the same for all Supernovae of this type, and this is the reason why these sources are also dubbed "standard candles."

Now, if the cosmic space in which both the Supernovae and the terrestrial observers are embedded were flat and static, of Euclidean type, then the received energy flux would be inversely proportional to the squared distance of the source. However, this is not what we observe: the received flux turns out to be *lower* than expected according to the rules of the Euclidean geometry, and such an anomaly increases as we consider more and more distant sources.

It was just on the grounds of similar observations that—almost a 100 years ago—it was discovered by Hubble and Humason the famous law describing the state of expansion of our Universe. If the Universe is expanding, in fact, the distance between source and observer is not a constant, but grows with time. The more the distant is a source, the stronger is the effect (since the emitted light takes more time to get to us): hence, as the distance increases, the received flux becomes fainter and fainter, and increasingly anomalous if compared to the situation of a static Euclidean space.

Triggered by this experimental result, and theoretically motivated by the gravitational theory of general relativity, the mid of the last century has witnessed the birth of the so-called standard cosmological model, which for many years has successfully explained and interpreted all observational data concerning the large scale structure of our present Universe. In this model, the geometric properties of the cosmic space are controlled by the distribution of matter and radiation according to Einstein's gravitational equations. Such equations tell us that the three-dimensional space must expand (thus explaining the weakening of the light flux received from the distant sources), and that the expansion must slow down with time (because it is damped by the mutual gravitational attraction among the matter sources).

It was thus a big surprise when, by repeating with increasing precision the measures of the received energy flux, and including more and more distant sources, an abnormal weakening of the expected flux was found: the observed flux is even fainter than predicted by the standard cosmological model (which takes into account the consequences of the cosmic expansion)!

The experimental data showing this effect, obtained from observations of Supernovae sources, were released for the first time towards the end of the Nineties and have been later presented in always more accurate compilations. According to these data, the weakening of the flux of the most distant sources is so large as to correspond to an *accelerated* growth in time of their distance (and not to a decelerated growth, as predicted by the standard model).

Hence, if we are not willing to modify neither the Einstein equations nor the usual properties of the effective gravitational interaction, we must unavoidably accept the idea of modifying the standard cosmological model. In particular, we have to seriously consider the possible presence, at a cosmic level, of some new type of gravitational source able to counteract the universal attraction due to matter (including all possible types of "dark" matter), and producing a net accelerated expansion, in agreement with the Supernovae data.

The Einstein equations, on the other hand, tell us that any type of cosmic stuff (or fluid) can determine the time variation of the expansion velocity in two ways: with a contribution proportional to its energy density and another one proportional to its pressure. Both these quantities are positive (or null) for the cosmic sources of the standard model (i.e., planets, stars, galaxies, interstellar dust, electromagnetic radiation, cosmic particles, dark matter, and so on). Hence, in order to produce the opposite gravitational effect, the new (non-standard) fluid component must be characterized either by a negative energy or by a negative pressure.

Excluding the (more exotic) case of negative energy—which could be problematic in a quantum context—we are then lead to conclude that the current evolution of our Universe should be mainly controlled by a cosmic element (dubbed "dark energy") with negative pressure.

At this point, we can officially open the competition to select the most suitable candidates to play the role of dark energy.

The simplest and more natural choice—adopted from the outset, and still valid—is the introduction of a cosmological constant Λ, corresponding to a dark-energy density independent of time and of the spatial position. According to the Einstein equations this particular type of energy must necessarily correspond to a negative pressure (equal in magnitude and opposite in sign to the energy density). In addition, the Einstein equations tell us that if the constant Λ represents the dominant source of gravity at the cosmic level, then the expansion of the Universe turns out to be automatically accelerated.

The introduction of dark energy in the form of a cosmological constant seems thus to provide a satisfactory description of the present cosmological state through a "minimal" modification of the standard model. However, two important problems unavoidably arise with the introduction of Λ.

The first problem concerns the very tiny value of the energy density associated to Λ: to agree with observations, in fact, this value has to be nearly 10^{122} times smaller than the value one would expect if Λ—as will be discussed in Sect. 3.3—would be determined by the properties of the quantum vacuum state. Really a huge discrepancy! On the other hand, if Λ has nothing to do with the properties of the vacuum, and corresponds to a (classical) arbitrary constant, then why its value is precisely the observed one? A value a bit smaller, and Λ would be totally negligible today (and the cosmic acceleration would be absent); a value a bit large, and the Universe could not exist in the form it takes today.

The second problem concerns the fact that the density of dark energy associated to Λ is constant in time, while the energy density of cosmic matter and radiation *decreases* in time. Today the dark-energy density is of the same order of magnitude as the matter density (the relative ratio is about 7 to 3). In the past, however, the matter density was dominant, while in the future the cosmological constant will become the dominant form of energy. Thus, why matter and dark energy have about the same density just at the present epoch? This seems to point at today as to a highly privileged epoch, and represents what is called the "coincidence problem."

In order to attempt a simultaneous resolution of all the above problems we are lead to consider the possibility that the dark-energy density is not a constant, but changes in time. An appropriate variation could explain why today this energy density is so small; also, it could ensure that the approximate equality of the energy densities is not a coincidence typical of the present epoch, but a permanent (or semi-permanent) property of the cosmic gravitational field, also valid in other epochs (if not for ever).

This interesting conjecture has stimulated the study of models in which dark energy is not represented by a constant but—more generally—by a field of forces present at a cosmic level, generically referred to as "quintessence." It is amusing to observe how this name makes explicit reference to the elusive fifth element (or ether) of the Aristotelian philosophy, an element long sought (but without success) by the medieval alchemists.

Dark energy—so far, at least—behaves indeed like a sort of "modern ether:" it has rather strange physical properties, it has been introduced *ad hoc* to explain a contradiction between observations and theoretical predictions, and nobody (up to now) has been able to obtain a direct proof of its existence.

3.2.1 The Cosmic "Quintessence"

A model of dark energy based on the existence of a new field (and a new particle) might seem to be much more reasonable and "realistic" than a model based on a cosmological constant—at least from the point of view of a theoretical physicist, used to represent fundamental interactions as fields and not as constants. Even in that case, however, it is not a simple task to justify with a compelling model the strange properties that the quintessence field should possess.

First of all, such a field should be characterized by an intrinsic potential energy which, irrespective of its initial value, changes in time in such a way as to follow (or, better, "to track") the time evolution of the matter energy density,[8] and to adapt its value to the matter energy density at a given epoch (for instance today).

In addition, in order to be effective on cosmological scales of distances, the range of the quintessence field has to be not smaller than the Hubble radius L_H. Hence, the particle associated to this field should have a mass not larger than $m \sim 1/L_H$, which is a tremendously small value, of the order of 10^{-33} eV (namely, about 10^{42} times smaller than the mass of a proton!). Given that the unavoidable effects occurring at the quantum level (in particular, the so-called

---

[8]This is what happens, for instance, in models where the behavior of the quintessence field is described by the so-called *tracker solutions*. See, e.g., the paper by Slatev, Wang, and Steinhardt [19].

"radiative corrections") tend to increase the mass of this particle—and thus to decrease the range of the quintessence—a consistent model should be able to avoid such a quantum "fattening" of the quintessential particle through some appropriate protection mechanism.

On the other hand, if the quintessence corresponds to a long range force, it will produce effects not only on the cosmic expansion but also on the gravitational interactions of ordinary macroscopic matter. To avoid introducing too large corrections at the macroscopic level, which would conflict with present experimental evidence, one must then require that the quintessential force acting on baryons—which provide the main contribution to the mass of ordinary matter—is much weaker than the usual (Newtonian or relativistic) gravitational force.

The interaction between quintessence and dark matter, on the contrary, cannot be negligible. Dark matter, indeed, represents today the dominant form of heavy (non-relativistic) matter present at a cosmic level,[9] while the quintessence field represents dark energy. To obtain a scenario where the time evolution of these two "dark" components is closely (and mutually) correlated, their reciprocal interaction must have a coupling strength not too different from the gravitational one (otherwise they would evolve independently under the action of the cosmic gravitational field, with no hope of solving the coincidence problem). But this implies that the action of the quintessence field is of non-universal type: it is weaker with baryons and stronger with the dark matter particles, hence violates the equivalence principle.

Given the many and different requirements listed above, difficult to be simultaneously implemented, it should be clear that the field of force representing the dark energy effects must possess highly uncommon properties.

It is possible, for instance, that its kinetic energy has a velocity dependence quite different from that of ordinary matter (i.e., not simply proportional to the square of the velocity): this is indeed what suggested by the so-called models of "kinetic quintessence," precisely invented to solve some of the problems posed by the introduction of dark energy.[10]

It is also possible, however, that there is no need of inventing *ad hoc* new fields and new models, and that the role of quintessence may be successfully played by some field already present in our models, and introduced for other reasons. For instance by the dilaton, a scalar partner of the graviton which is necessarily required by string theory.

The dilaton, in fact, is characterized by a potential energy which goes rapidly to zero when the field is weak,[11] and which may have a more complicated structure in

---

[9]Current observations tell us that among the total matter present in our Universe, and composed of heavy, non-relativistic particles, there is a fraction of baryonic matter corresponding to about 10 % of the total. All the rest is composed of dark matter particles, of the same type as those particles that modify the rotation velocity of the stars inside a galaxy.

[10]See for instance the paper by Armendariz-Picon, Mukhanov, and Steinhardt [20].

[11]More precisely, when the exponential of the dilaton field (which controls the strength of all interactions) is much smaller than one.

Fig. 3.1 A possible example of dilaton potential. The strength of the interactions grows as dilaton field ϕ is growing. The *dashed curves* reflect our current ignorance of the strong-field regime, where the dilaton potential might be decreasing or might explode to infinity

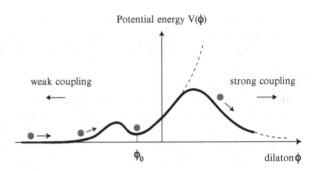

the opposite limit in which the field is strong (see Fig. 3.1). In the weak field regime the corrections due to quantum effects can be safely neglected, while, for strong enough fields, they must be necessarily included.

In many cosmological models suggested by string theory (see in particular the discussion of Chap. 6), the intensity of the dilaton field is initially very low, but the field grows in time and is thus doomed to reach the region of oscillating potential illustrated in Fig. 3.1. At that point there are two possibilities.

A first possibility[12] is that the dilaton gets trapped and stabilized in a minimum of the potential, for instance at the value ϕ_0 of Fig. 3.1. This can easily occur in the past of our Universe, when the dominant form of cosmic energy was the electromagnetic radiation. The radiation, in fact, has only a very weak influence on the dilaton evolution, especially in the regime where the field is weak and the quantum corrections are not important.

However, even if the dilaton is "sleeping" at the bottom of its potential well during the radiation-dominated era, when the Universe becomes matter dominated the dilaton, driven by the evolution of the matter energy density, "weaks up" and tends to escape from its equilibrium position. The shift away from the minimum can be avoided only if the minimal potential energy, $V(\phi_0)$, belongs to an appropriate range of values and if, in addition, the dilaton coupling to dark matter is not too strong. If such conditions are satisfied then the dilaton keeps stable, and the Universe evolve towards a final configuration dominated by the constant value of the dilaton potential energy, acting as an effective cosmological constant $\Lambda = V(\phi_0)$.

In this scenario the coincidence problem is not completely solved but, at least, turns out to be alleviated: in fact, not all values are allowed for the minimum of the dilaton potential, but only the limited range of values compatible with a final dilaton-dominated regime. This means that the approximate equality between matter density and dark energy density cannot occur in principle at any epoch, but only at those epochs included in a special time interval, whose precise limits are determined by the properties of the dilaton interactions.

---

[12] A concrete example has been discussed in a paper I wrote in 2001 [21].

As a consequence, our epoch is not the "most special" of all the infinite epochs of our Universe, but only the most special within the restricted interval of epochs allowed by the dilaton dynamics.

There is, however, also a second possibility to be considered: the case in which the initial velocity of the dilaton is large enough to pass through the intermediate region of the potential—the region with the maxima and minima illustrated in Fig. 3.1—and definitely enter the regime of strong interactions. In that regime the quantum corrections provide decisive contributions, and the role played by the asymptotic behavior of the potential, in that case, is crucial.

If the potential (and thus the strength of the interactions) keeps growing in an unbounded way, then there is practically no hope to extract reliable predictions. If, on the contrary, even including all quantum corrections, the strength of the dilaton interactions tends to saturate to small and negligible values for the baryons, and to larger values for the dark matter particles, then there is a chance to obtain a viable scenario able to explain (or to remove) the so-called cosmic coincidence.

This is indeed what happens in a class of models based on string theory,[13] where the potential energy $V(\phi)$ goes exponentially to zero when the dilaton field becomes arbitrarily large (see Fig. 3.1). In that case the dilaton is never stabilized to some constant value, and thus its energy—representing the cosmological dark energy—is always a mixture of kinetic and potential energy. Our Universe is lead to its present state after a long preliminary phase which has the virtue of "adapting" to each other the values of the different forms of energy densities.

The initial, preliminary phase starts as soon as dark matter becomes the dominant form of cosmic energy, overcoming the contribution of the radiation background.[14] At that time the dilaton potential energy is still negligible with respect to the dilaton kinetic energy, which, in turn, is negligible with respect to the matter energy density. Nevertheless, the coupling between dilaton and dark matter is strong enough to force the dilaton to have a time evolution which closely follows the evolution of matter. This phase is called indeed "dragging phase," since the dilaton is "dragged" by the dark-matter energy density.

The next phase, called "freezing phase," is triggered by the potential energy of the dilaton as soon as it becomes comparable with its kinetic energy and with the energy density of matter. Thereafter, all ratios between the various dominant contribution to the total energy density (i.e., between the matter and the dilaton kinetic contribution, between the matter and the dilaton potential contribution, and so on) are frozen at constant values of the order of unity, and so remain throughout the future evolution.

It is important to stress that, during the freezing phase, all dominant components of the total energy density have the same time dependence (as their relative ratio

---

[13] See for instance the paper by Gasperini, Piazza, and Veneziano [22].

[14] This occurs when the Universe crosses the so-called equality epoch, characterized by a temperature about 10,000 larger than the present temperature of 2.7 K. We can say, more precisely, that the Universe becomes matter dominated when the radiation temperature falls below 14,700 K (see, e.g., [2]).

is constant in time), but such a time dependence is different from that typical of dark matter in the absence of the dilaton, and from that typical of the dilaton in the absence of dark matter.

The energy density of dark matter, if alone, would be rapidly diluted in time like the inverse of the expanding volume and would sustain a Universe in decelerated expansion. The dilaton energy density, if alone, would tend to be dominated by its potential and thus to behave like an effective cosmological constant, driving the Universe towards a phase of accelerated expansion. The model introduced above, on the contrary, predicts for all the cosmic components a time dependence of intermediate type: the energy density of the dilaton and of dark matter is both diluted in time, but at a rate slow enough to produce a phase of accelerated expansion.

If this scenario is right, we have entered (maybe only recently) the final freezing phase, and the ratio between the dark-energy and dark-matter components that we are currently observing (about 7 to 3) is going to change very little (if not at all) in the future cosmological eras. Hence, our epoch is not privileged at all in comparison with the endless sequence of later epochs, and we are not experiencing any kind of "coincidence."

The conclusion is appealing, but there is a question that we should ask: are there observations that could experimentally confirm (or disprove), either directly or indirectly, the proposed scenario?

Let us put aside, for the moment, the possibility of observing the future constancy (or the future variation) of the ratio dark-energy to dark-matter density, which seems to require too long observation times and/or observations of exceeding precision. Then we are left with two possibilities: trying to determine, experimentally, (1) the behavior in time of the ratio between dark and baryonic matter during the freezing phase, and (2) the epoch of the onset of the accelerated regime.

In the freezing phase, in fact, the time dependence of the dark-matter density changes so as to adapt itself to the dilaton evolution: in particular, the dilution rate of dark matter slows down due to the presence of the dilaton. Baryon matter, on the contrary, is fully decoupled from the dilaton and keeps a faster dilution rate. As a consequence, the fraction of baryon to dark-matter density observed today is smaller than in the past, and will be always smaller as time goes on. This is not the case in models where dark energy and dark matter are fully decoupled (like those based on a cosmological constant).

Any experimental indication about the fraction of baryon matter present in some past epoch (for instance, at the equality epoch), compared with the current value of the baryon fraction, could thus give us important information about the validity of the above dilaton scenario.

Another possibility to test such a scenario relies on the fact that, in the presence of a strong coupling between dark energy and dark matter, the Universe may start accelerating in epochs far more remote than those expected in models with a weaker coupling.[15] Unfortunately, however, the localization in time of the beginning of the

---

[15] As shown in a paper by Amendola, Gasperini, and Piazza [23].

accelerated era is currently known with a large uncertainty, and will be possibly determined with a better accuracy by future experiments only by including more and more distant Supernovae, or using other types of cosmic sources (for instance gamma ray sources), able to further extend our observational range far away in space and back in time.

So, are we experiencing today a cosmological phase dominated by the dilaton, or by some other "quintessence" field which plays the role of dark energy, and drives the current accelerated expansion of our Universe? The answer will come, we hope, from future observations. In the meantime we can (and should) consider other possible explanations (or interpretations) of dark energy, like those based on the vacuum energy density that we will introduce in the following section.

3.3   The Fluctuations of the Vacuum Energy

What is the vacuum according to modern physics? It is the state of minimum energy. And what is the minimum energy of the cosmic space? Imagine we remove all forms of matter and radiation: the space would still contain an intrinsic energy of quantum origin, provided by the uncertainty principle. What is the value and the form of this residual energy? Could it play the role of the mysterious dark energy?

In order to answer these questions we should recall that all physical fields, representing either matter or interactions, can be exactly set to zero only in a classical and macroscopic context: they cannot be completely canceled at the microscopic level because of the existing quantum effects. For instance, we can arrange a situation where the classic part of the electric field is zero in a given region of space, but we cannot avoid that—in the same region—the total electric field locally deviates from zero with extremely small, fast, and unpredictable "fluctuations" of quantum origin, different from one point to another and from one instant to another.

In particular, we can think of the quantum component of a field as a superposition of a great number of very small waves, each of them with a different wavelength λ and a minimal energy—called "zero-point" energy—which is constant, different from zero and inversely proportional to λ. The space, even if (classically) empty, cannot avoid receiving the energetic contribution of all such tiny quantum "ripples."

Should we include all such waves in the computation of the quantum vacuum energy? Yes, in principle; in practice, however, it is enough to include only those with a wavelength smaller than the maximal distance scale (or with a period smaller than the maximal time scale) that we are able to measure. For the other waves we would not be able to appreciate the spatial (o temporal) variation of their amplitude, and the corresponding energy—which is precisely determined by such variations— would be zero to all practical purposes.

By adding the minimal energy E_λ of all waves (which grows as the wavelength decreases, since $E_\lambda \sim 1/\lambda$), we find that the value of the total energy is controlled by the smallest waves we have included, and that such a value goes to infinity as λ goes to zero! This is true for all fundamental fields that exist in Nature, and that behave in the quantum regime as a superposition of waves of arbitrarily small length.

So, should we conclude that the vacuum energy is infinite? There are two possible ways out to evade this conclusion.

A first possibility[16] is "to cut" off the contributions of all wavelengths shorter than a given "minimal" length scale L (either because at distances smaller than L the model we are using is no longer applicable, or because the model itself tell us that distances smaller than L do not make sense).

In that case the contribution of the shortest wavelength $\lambda = L$ is the one that controls the total vacuum energy E, hence this energy turns out to be inversely proportional to the considered minimal length: $E \sim 1/L$ (the contribution of longer waves is negligible compared to this). The associated energy density Λ, i.e., the energy per unit volume, is then inversely proportional to the fourth power of the minimal length scale: $\Lambda \sim E/L^3 \sim 1/L^4$.

In such a context, the quantum energy density of the vacuum turns out to be finite. Also, it turns out to be constant (if L is constant), and thus equivalent in all respects to an effective cosmological constant, precisely as required to simulate the observed dark energy effects. However, the constant obtained in this way is very large, since the distance scale where we can impose the cutting of the smallest quantum ripples is certainly very small.[17]

Consider, for instance, the gravitational field: the model we are currently using (the theory of general relativity) is expected to be valid up to the distances of the order of the Planck length, $L_P \sim 10^{-33}$ cm. The vacuum energy density associated to this scale, $\Lambda_P \sim 1/L_P^4$, is very huge, and can be expressed as $\Lambda_P \sim (10^{19}\,\text{GeV})^4$ (we should recall that, as also pointed out in Chap. 2, the energy scale corresponding to the Planck mass $M_P = 1/L_P$ is 10^{19} GeV). The result is not so much different if, instead of L_P, we take as limiting distance the length scale L_S typical of a quantized string: in fact, string models unifying all interactions suggest for L_S the value $L_S \sim 10\,L_P$, very similar to the value of the Planck length.[18]

A second way to avoid an infinite vacuum energy is to assume that the quantum contributions of all fields, although infinite if taken individually, can cancel each other to give an overall result which is zero, or at least finite.

This possibility (probably more realistic, and certainly better motivated than the previous one) is based on the fact that the zero-point energies of the fermionic fields have the opposite sign of the bosonic zero-point energies. In models where there is the same number of boson and fermion fields (with the same masses), and where

---

[16]The removal of the shortest wavelengths approaching zero is a procedure usually applied in many physical situations. We say, in that case, that we have introduced an "ultraviolet cutoff."

[17]See for instance the review paper by Weinberg [24].

[18]As we shall see, however, there are possible exceptions due to the presence of extra spatial dimensions.

supersymmetry is implemented exactly, the vacuum energy can thus be identically zero.[19]

Our Universe has probably experienced a phase characterized by exact supersymmetry in its very remote past, but, for sure, its present state is not exactly supersymmetric. Otherwise, all particles we are presently observing would be organized in supersymmetric multiplets of particles of equal masses. This means, for instance, that, together with the photon (spin 1) and the graviton (spin 2), which are massless, we should also observe a photino (spin 1/2) and a gravitino (spin 3/2) also massless, hence easy to produce and to detect even at low energies.

This is not what happens, actually, hence our present Universe is not exactly supersymmetric, and its vacuum energy is nonzero. The Universe, however, could be in a state of broken supersymmetry, namely a state in which the exchange symmetry between bosons and fermions does not apply exactly but is implemented in approximated form, because the various particles have masses very different from those of their supersymmetric partners.

The breaking of supersymmetry, in fact, induces among the particles of the same supersymmetric multiplet a mass difference of the same order as the energy scale at which the breaking occurs (let us call M_{SUSY} this energy, and denote with $L_{\mathrm{SUSY}} = 1/M_{\mathrm{SUSY}}$ the associated length scale). The fact that we are not currently observing the partners predicted by supersymmetry could mean, simply, that M_{SUSY} is too large compared to the energy scales presently accessible to our experiments. This means, in practice, $M_{\mathrm{SUSY}} \gtrsim 1\,\mathrm{TeV} = 10^3\,\mathrm{GeV}$.

If this is the case, namely if supersymmetry is restored in its exact form only at energies larger than M_{SUSY} (i.e., at distances smaller than L_{SUSY}), then the exact cancelation between the bosonic and fermionic zero-point energies is effective only for wavelengths short enough to belong to the supersymmetric regime ($\lambda < L_{\mathrm{SUSY}}$). For the other wavelengths, on the contrary, there is no cancelation: the net result is thus a vacuum energy density which is finite, nonzero, and inversely proportional to L_{SUSY}.

The corresponding energy density is again equivalent to an effective cosmological constant that, in this case, satisfies the condition $\Lambda_{\mathrm{SUSY}} \sim 1/L_{\mathrm{SUSY}}^4 \gtrsim (10^3\,\mathrm{GeV})^4$. So, if Nature is (at least approximately) supersymmetric, then the value of the vacuum energy density can be, if not vanishing, at least much smaller than the empirical value $\Lambda_P = 1/L_P^4$ associated to the Planck length. Unfortunately, however, even this result is too large to be phenomenologically viable, and amazingly far from the value required to simulate the observed dark energy effects.

In fact, the cosmic acceleration we are currently observing, if interpreted in the context of Einstein's equations with a cosmological constant Λ, requires for Λ a

---

[19]As shown for the first time by Zumino [25]. They are also supersymmetric models where the vacuum energy density is constant and negative, but they do not seem to be compatible with a supersymmetric description of all interactions based on string theory, as discussed, for instance, by Witten [26].

value that can be expressed in terms of the Planck length and of the Hubble length as $\Lambda \sim 1/L_P^2 L_H^2$. Such a value is 122 orders of magnitude[20] smaller than Λ_P, and at least 54 orders of magnitude smaller than Λ_{SUSY}. How to explain such a terrible discrepancy?

Let us notice that, even if the cosmic acceleration were absent (or negligible), it would remain to explain why the huge energy density Λ_P or Λ_{SUSY} associated to the quantum fluctuations does not produce any observable effect. What happened to Λ_P or Λ_{SUSY}? This is definitely one of the biggest problems of modern physics: there are many proposed solutions, but none of them is currently believed to be definitive.

Given the presence of the cosmic acceleration, the problem to be solved is actually twofold: we have to not only eliminate the big energy contribution arising from the short-wavelength part of the spectrum of quantum fluctuations, but also to justify the extremely small (but nonzero!) contribution that seems to survive and to be responsible for the observed effects on large scales. Is it possible to find satisfactory solutions to these problems in the context of the fluctuations of the vacuum energy?

Adopting a phenomenological approach, and considering an arbitrary distance scale L, we could assume (as a working hypothesis) that, in general, there are two possible contributions to the vacuum energy. Let us suppose, in particular, that besides the usual contribution inversely proportional to the distance, $E \sim 1/L$, there is the possibility, at large enough scales, of an additional contribution which turns out to be *directly* proportional to the distance scale L, and which can then be written[21] as $E \sim L/L_P^2$.

This would give, at all sufficiently large scales, a new term corresponding to an energy density $\Lambda \sim E/L^3 \sim 1/L_P^2 L^2$. For a distance of the order of the Hubble radius, $L = L_H$, we would obtain the value $\Lambda \sim 1/L_P^2 L_H^2$, which is exactly the cosmological constant needed to explain the cosmic acceleration!

Assuming that we can neglect, for some reasons, the small distance contribution (generating the energy density $1/L_P^4$ or $1/L_{\mathrm{SUSY}}^4$), the above hypothesis could thus explain the effects observed on a cosmological scale as a consequence of the cosmic fluctuations of the vacuum energy. But then, how to justify the presence of the new contribution directly proportional to L?

Maybe through the so-called duality symmetry typical of string theory (see Sect. 5.5), which can make physically indistinguishable a length scale and its inverse. Or maybe through quantum fluctuations leading to the spontaneous production of branes, thanks to the energy of a special antisymmetric field of forces

---

[20]In order to obtain a quantitative estimate, and to compare the different results, we have to use the numerical values of the reference length scales L_P, L_{SUSY} and L_H. They are given, respectively, by $L_P \sim 10^{-33}$ cm, $L_{\mathrm{SUSY}} \sim 10^{-16}$ cm, and $L_H \sim 10^{28}$ cm.

[21]The proportionality factor connecting E and L has to be the inverse of a squared length, for dimensional reasons. Here, as we are mainly referring to the gravitational interactions, we have chosen the fundamental Planck length for the definition of the proportionally factor $1/L_P^2$.

predicted by supersymmetry.[22] Or maybe through the hypothesis[23] that the vacuum energy fluctuates from one point of space to another following a Poissonian distribution,[24] and that it is localized inside microscopic cells (or elementary "atoms of space") of Planck radius L_P.

In this last case, in fact, if we take a region of space geometrically confined within a distance of radius L, and containing a number N of these small cells, we obtain (for the properties of the Poisson distribution) that the energy fluctuations are of the order of $E \sim \sqrt{N}/L_P$. Assuming that the relevant number of cells, for this mechanism, is the number of cells present *on the surface* (and not inside the volume) of the considered region of space, we find $N \sim L^2/L_P^2$, and we eventually obtain that the overall energy of such fluctuations is directly proportional to the given distance, $E \sim L/L_P^2$, exactly as proposed before.

But, even if this is true, we have still to dispose of the huge energy of the vacuum fluctuations arising from their contribution inversely proportional to the distance. Otherwise the contribution of the shortest length scales would be the dominant one, leading to unacceptably high levels of the vacuum energy density.

To this purpose there are various proposals. Among the most radical ones we can mention, for instance, the assumption that such a contribution is "gravitationally neutral," and then has no effect on the gravitational field present on cosmological scales; or the assumption that the total contribution of the shortest distance scales is intrinsically zero because of an exact cancelation due to a change of sign in the energy of the quantum fluctuations, below a given scale, when λ tends to zero.[25] But if we prefer to avoid these (and other) exotic possibilities, mostly introduced *ad hoc* and often lacking of sound physical motivations, the extra dimensions—and, in particular, the brane model of space–time—come again to our rescue.

In fact, if the space in which we live is a three-brane, i.e., a three-dimensional slice of a higher-dimensional external space, it is quite possible that the big vacuum energy localized on our brane affects the spatial geometry only outside the brane, bending the extra dimensions according to Einstein's equations but leaving our space flat (as in the absence of vacuum energy). A similar effect would not be unusual, after all: it is well known, for instance, that the gravitational field generated by the cosmic sources (stars, galaxies, and so on) bends our space–time but leaves almost flat the geometry of the three-dimensional space.[26]

---

[22]See, e.g., the discussion by Bousso [27].

[23]See, e.g., the paper by Padmanabhan [28].

[24]The statistical distribution of Poisson is characterized by the property that the variance (or mean square deviation) of a distribution coincides with its mean value.

[25]There is also a recent proposal of computing the quantum energy of the vacuum by including, for any positive energy state, the contribution of a corresponding negative energy state, according to the so-called procedure of time-symmetric quantization [29]. The model, however, is affected by formal problems when trying to describe interacting fields.

[26]Current observations tell us that the mean curvature of the three-dimensional space, at the cosmic level, is equal to zero with an accuracy of one percent. See for instance the official site of the

The interesting aspect of this "neutralization" of the gravitational effects of the vacuum on the brane is that we can never completely eliminate, in this way, the vacuum energy of the brane arising from the process of supersymmetry breaking. A (small) fraction of the energy density Λ_{SUSY}, offloaded along the extra dimensions, necessarily comes back onto the brane representing our three-dimensional space.[27] This effect allows us to obtain precious information on the energy scale at which supersymmetry is broken (or restored), and also, indirectly, on the brane-world model that we are using.

In fact, if the energy density $\Lambda_{\text{SUSY}} \sim 1/L_{\text{SUSY}}^4$ affects the geometry of the extra dimensions, then, according to Einstein's equation, necessarily induces in the space outside the brane a local curvature radius $r \sim L_{\text{SUSY}}^2/L_P$. Such a curvature breaks supersymmetry in the extra dimensions,[28] and the energy scale of this breaking turns out to be inversely proportional to the induced radius of curvature r. The breaking of supersymmetry, on the other hand, induces in the higher-dimensional "bulk" space an energy density which is of the order of $1/r^4 \sim L_P^4/L_{\text{SUSY}}^8$. Our three-dimensional brane, being embedded in that space, cannot avoid to absorb, in its turn, such an intrinsic vacuum energy, and thus unavoidably acquires an effective cosmological constant $\Lambda \sim 1/r^4 \sim \Lambda_P(L_P/L_{\text{SUSY}})^8$.

Given that $L_P \ll L_{\text{SUSY}}$, such a residual energy density turns out to be much smaller than the Planckian value Λ_P, and also much smaller than the value Λ_{SUSY}, originally produced by the breaking of supersymmetry directly on the brane. Hence, it could play the sought role of dark energy and be responsible for the current state of accelerated expansion. In any case, and above all, it has to be small enough to avoid conflicting with the present observations.

This implies, in particular, that the ratio $(L_P/L_{\text{SUSY}})^8 = (M_{\text{SUSY}}/M_P)^8$ must be smaller than (or at most equal to) about 10^{-122}, a very small number expressing in Planck units, the currently allowed value of the cosmological constant. By imposing this condition we then obtain[29] $M_{\text{SUSY}} \lesssim 10 \, \text{TeV}$: namely, a scale of supersymmetry breaking that could be within the reach of the experiments carried out in the accelerator LHC at CERN in Geneva!

These experiments, planned to study the collisions of particles of very high energy, should in fact be able to detect (or to exclude) the presence of supersymmetric particles up to energies of about $14 \, \text{TeV}$. Hence, they could also indirectly confirm (or exclude) the possibility that the current acceleration of our Universe is due—at least partially—to the quantum energy of the vacuum, originated as a consequence of supersymmetry breaking and "diluted" in the extra dimensions external to our macroscopic three-dimensional space.

Particle Data Group at http://pdg.lbl.gov, containing an updated compilation of all relevant data for particle physics, astrophysics, and cosmology.

[27] This effect has been discussed in a paper I wrote recently [30].

[28] See for instance the paper by Antoniadis, Dudas, and Sagnotti [31].

[29] To obtain this result we must insert the numerical value of the Planck mass, $M_P \sim 10^{19} \, \text{GeV}$, and recall that $1 \, \text{TeV} = 10^3 \, \text{GeV}$.

Chapter 4
Space, Time, and Space–Time

Let me start this chapter with a somewhat "philosophic" introduction, concerning the so-called reality of physical models.

As a physicist, I am a bit reluctant to deal with similar topics which—strictly speaking—are beyond my professional expertise. However, when we are dealing with the notions of space and time, which are at the ground of the physical description of Nature that we have so far (and so laboriously) achieved, I think we cannot avoid making some observations and asking ourselves a few uneasy questions.

For instance, even if a physicist does not frequently talk about this (not with colleagues, in particular!), I believe that every physicist, sooner or later, has wondered: what is time? what is space?

These are the questions of mainly philosophic type, which probably do not make much sense in a physical context. Or, better, are questions allowing a circular reply. For instance, to the question "What is time?" a physicist could answer: "It is a parameter which describes motion." And what is motion? "It is an evolution in time." And so we are back to the starting point.

Maybe the answer to the above question does not exist, or maybe the question, as we have expressed it, is not well posed. However, trying anyway to give an answer that honestly reflects my personal opinion about these issues, I would say that time and space (and their geometric fusion, the so-called space–time used in the context of the relativistic theories) are models produced by our mind, well suited to describe the physical phenomena that we know so far. But they are no more "real," for instance, of the abstract Hilbert space used by quantum theories to describe the propagation of the probability waves.

Notions like "here," "there," "now," "yesterday" are a fruit of our mind which has invented these concepts as an output to the received sensorial stimuli, and has produced a model of space–time adapted to a quantitative description of the real world around us. However, the true essence of such a "reality," in my opinion, is elusive.

Every time an old model is replaced by a more refined one, which provides a best fit of the experimental results—as, for instance, in the case of the absolute

M. Gasperini, *Gravity, Strings and Particles*, DOI 10.1007/978-3-319-00599-7_4,
© Springer International Publishing Switzerland 2014

time of Newton's theory replaced by Einstein's relative time—we believe we have
discovered how Nature is made and how is working: actually, however, we have
only invented a more faithful and effective way of interpreting and organizing our
sensorial experience. We have more closely approached the reality of things, but we
have not "really understood" their essence, nor (in my opinion) we will ever be able
to understand this completely.

4.1   Maybe the Past Can Change?

I do not know whether most of the readers of this book believe that our own future
is already written. Personally, I don't think so. In the same way, just as the future is
not already written, maybe even the past is not unchanging. Some (or many) of us
are willing to accept the idea that the future can change according to our actions, but
very few of us, I believe, have ever thought about the possibility that even the past
may change.

In order to better clarify what I mean, let me start with some elementary
observations about the concepts of time and motion.

The concept of motion (namely, of change of position or, more generally, of
change of state of a system) is certainly one of the basic pillars for the description
of Nature provided by physics. We may say, in a sense, that the whole of physics is
nothing more than a collection of laws aiming at the prediction of the time evolution
of the world around us, starting from the evolution in time (i.e., from the motion) of
its fundamental microscopic components. But what do we mean, exactly, when we
say that an object is moving?

Let us consider a simple example concerning the classical dynamics of a point-
like particle. If we say that a particle is moving we simply mean that such a particle,
observed at a spatial position x_1 at the time t_1, is observed at a different spatial
position x_2 at the time t_2 (see Fig. 4.1). This is an operative definition of the concept
of motion, based on experimental observations, about which everyone will certainly
agree. Such a definition, however, is open to two possible interpretations.

A first, more conventional interpretation is that the particle has actually changed
its position from the point x_1 to the point x_2, proceeding along a space–time
trajectory (commonly named "world-line") which describes, step by step, instant
by instant, its localization in space and time. In that case the particle (if exactly
point-like) at the time t_1 is located in x_1 and nowhere else, at the time t_2 is located
in x_2 and nowhere else; and so on.[1]

---

[1] The times t_1, t_2 and the positions x_1, x_2 are obviously referred to the clock and to the coordinate
system of a given observer. For a different observer the corresponding values of times and of the
spatial coordinates could be different. What matters, in this example, is only the variation of the
position with time.

Fig. 4.1 Simple example of space–time trajectory describing the motion of a point-like object along the x axis

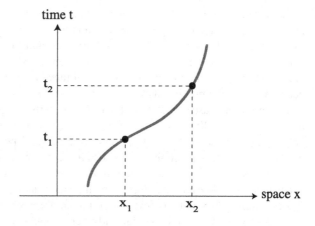

A second interpretation (which might seem strange, but which is also possible) is that the particle, even if point-like for what concerns the spatial extension, has a (possibly finite) temporal extension that coincides with the trajectory illustrated in Fig. 4.1. This means, in other words, that we take as a real physical object associated to the particle the set of all points/events located on its world-line: a sort of thin space–time "wire," whose different sections, taken at different times (t_1, t_2, etc.), are placed at different spatial positions (x_1, x_2, etc.). In such a case, why should we say that there is a point-like object which has moved from the point x_1 to the point x_2?

Maybe because "we" ourselves are moving along the time axis (and the word "we" can be interpreted as our consciousness, or perception of the world, or—more concretely—our system of reference, understood as our physical capability of interacting and exchanging information within the three-dimensional space in which we are embedded). If this were the case, and if—for the existing limits on the speed of propagation of information—we could observe at the time t_1 only the point-like section of the "wire-particle" located in x_1, at the time t_2 only the section located in x_2, and so on, we would still say that the particle has moved from x_1 to x_2.

What is the difference between the two interpretations given above? An important conceptual difference which concerns our way of understanding the space–time model used in physics.

In the first case we are imaging space–time to be a non-static, incomplete entity, continuously evolving and growing,[2] in which what we call the "future" gradually emerges point by point and instant by instant, while—simultaneously—the "past" crystalizes into a definite configuration, recorded by our senses and by our instruments.[3]

---

[2]This scenario is also called "Emerging Block Universe," see for instance the contribution of G.F. R. Ellis to the book *Springer Handbook of Spacetime* [32].

[3]Obviously, what is meant for the past and future may vary according to the considered observer, and may also depend on the local geometric properties characterizing the considered portion of space–time.

In the previous example, in particular, the space–time continuously "concretize" along the world-line of the given particle, and this locally occurs along the evolution trajectories of all bodies present in our Universe, in all regions of space. In such a context, the past events (along a world-line) are fixed and immutable, the future ones are indeterminate and do not exist until they "happen," instant by instant, and become part of the past events.

In the second case, instead, all the different parts of the so-called wire-particle are always physically present in the space–time structure: the section t_1 exists both before and after an observer has recorded its spatial position x_1, and so on for the section t_2 at the position x_2 and for all the other sections. This is true for all components of our Universe; hence, in this context, the whole space–time always exists in its complete and final form, which includes all past and future events[4] (even if space–time only gradually reveals itself to an observer exploring its structure).

This second interpretation—currently very popular among physicists[5]—would seem to clash with the philosophical request of "free will." If all particles, and with them the whole physical world around us (including ourselves, since we are made of particles) are represented by objects extended in time (towards the past as well as the future), then it would seem that everything is already fixed and rigidly preset. The past cannot be different from what we have recorded, but also the future is already "written," as the space–time trajectories of all particles are fully extended in time: we can only gradually discover their future evolution (as our reference system flows along the time direction), without being able to influence such evolution.

This would be true if what we have called wire-particles were rigid objects, statically embedded in space–time. These space–time "wires", however, could vibrate, oscillate in the (x, t) plane of Fig. 4.1 (actually, they should do that because of the quantum fluctuations); also, interacting among themselves as a consequence of such fluctuations, they could even change shape, position, length, thus producing a matter and an energy distribution continuously changing, at the microscopic level, both towards the past and towards the future. Using a simple analogy, we could imagine the space–time filled with wire-particles as a meadow full of grass blades, ruffled by wind: seen as a whole it is a single, solid, and motionless block, but the grass blades of the meadows are continuously rippling, intersecting, and intertwining one another, always creating different configurations.

Such intrinsic fluctuations of the fabric of space–time are of quantum and microscopic nature. It is not impossible, however, that they may occasionally accumulate in a coherent way so as to generate macroscopic variations, in principle everywhere, hence, in particular, even in regions classified as "future" or "past" by a given observer. In that case, not only the future would be variable, unpredictable, influenced by previous events, but also the past would not be unchanging. If so, could we check this effect in some way, directly or indirectly?

---

[4] This scenario is also called "Block Universe", or frozen Universe. See for instance P. C. W. Davies [33].

[5] See also the discussion in the book by J. B. Barbour [34].

The answer would seem to be negative. We cannot go back in time, of course, to see if something has changed (it is forbidden by the laws of relativistic dynamics, because we should travel faster than light). Other observers might have experienced a past different from the one we ourselves have recorded, but, even in that case, they could not inform us of the differences (still because of the relativistic laws[6]).

Could we ourselves receive signals from the past, informing us that something has changed? In principle yes; however they would not be interpreted as signs of changes: being received *after* the initial information, they will be emitted by different and *farther* spatial positions, and hence they would not cancel previous information but would simply add as a further characterization of the past configuration we are recording.

So, we might be tempted to conclude that the past may change but, unfortunately (or thankfully!), only "unbeknown to us" and, anyway, only in a physically unimportant way. However, maybe things are not quite so.

A recent approach to the cosmological constant and to the dark energy problem, in fact, introduces a model[7] where the vacuum energy density Λ is always constant, but the precise value of Λ entering the (classical) gravitational equations may be different at different epochs, as it depends on "when" Λ is computed!

If we compute it today, at the time t_0, we find for instance a constant value Λ_0, and we can obtain a realistic model of Universe where today the contribution of the cosmological constant is dominant, while in the past, for instance at the epoch $t_1 < t_0$, its contribution was negligible. But if we compute it in the past, at the epoch t_1, we find a different constant value Λ_1, and a different model of Universe where the cosmological constant turns out to be dominant at the time t_1, and negligible at earlier epochs $t < t_1$. And so on.

Hence, in such a context the past—at least at the cosmological level—is not unchanging, in the sense that (literally quoting from the paper by Barrow and Show) *"we do not see the past as an observer in the past would have seen it"* [35]. The scenario is similar to the case of a space–time globally blocked but locally fluctuating illustrated before, with one main difference: the fluctuating classical trajectories, in this second case, are not those describing the evolution of a single particle, but those describing the global evolution of a cosmological system characterized by a constant energy density Λ.

The fact that such a scenario can be formulated in a rigorous mathematical way, consistent with the laws of relativistic and quantum physics, and—most important—the fact that it is able to make specific predictions which can be tested and possibly falsified by future precision experiments [35], clearly show that the idea of a "changing past" can be seriously (and fruitfully) considered in the context of modern physics.

---

[6]Such observers, for kinematic or geometric reasons, would be "causally disconnected" from us, namely they could not send us information through signals that do not exceed the speed of light.

[7]Discussed in a recent paper by J. D. Barrow and D. J. Shaw [35].

4.1.1 Time and Memory

Let us take seriously, for a moment, the idea introduced in the previous section, i.e., the possibility that the past "still exists," in a form not necessarily identical to the one we have experienced, and that the future "already exists," in a form not necessarily identical to the one we will see. This is so because Nature (or the Universe, or the physical reality, let us call it as we prefer) is fully expanded not only in space but also in time, and is in a state of continuous evolution, motion, transformation. In such a case, it follows that there is a close link between the physical quantity that we call "time" and the property of our brain that we call "memory."

It must be stressed in fact that physics—which is a men's construction—describes Nature as it is perceived by human beings, and thus provides a description heavily anchored to our senses (as well as to our instruments), and strongly correlated to the workings of our brain. If we had no memory, in particular, we would live in a static time, or eternal present. It is precisely our memory that is the basis for the concept of passing time, for the distinction between past and future and for the description of motion.

Let us try to imagine, for instance, a race of alien creatures without memory, perpetually living in a lethargy state allowing them to wake up only for a brief instant every night at midnight. During the waking moment their astronomers scan the sky and record the location of the celestial bodies in their books (for instance, the position of the moon with respect to the stars), and then fall back asleep immediately. The following night, observing the moon in a new position, and comparing it with what shown in their books, they could conclude—being without memory—that the notes reported in their books (recorded by who knows who!) certainly refers to a static Universe which is not their Universe, as it contains celestial bodies in completely different positions.

Our memory (and only our memory) would lead us to conclude, instead, that the two different records refer to the same Universe, observed at different times. Which is the right conclusion? Or maybe are both conclusions possible and acceptable? Are we entering, instant by instant, a (continuous or probably discrete) series of static Universes, each of which is infinitesimally different from the adjacent one? Is our concept of motion only an illusion due to this passage through the various Universes? Is it possible that different observers may pass through sequences of different static Universes? Will we ever be able to free ourselves from the conditionings of our memory, and to develop a higher-level physical description allowing us, in particular, to get a better understanding of the meaning of time?

All these questions remain (for the moment) with no answer. We should recall, however, that only by overcoming the conditionings of our senses we have been able to generalize our physical models imagining, for instance, the existence of the extra spatial dimensions, which our senses do not conceive and our instruments do not (still) detect.

4.1.2   Time: An Intrinsic Property of All Bodies?

In the context of modern relativistic physics we are used to consider space and time not as separate entities; we adopt in fact a model (introduced more than one century ago by Minkowski) where the two concepts are fused into a single geometric representation: the "space–time" manifold.

The use of such a unified geometric representation has enormously simplified the description of the relativistic processes and the formulation of a satisfactory gravitational theory (general relativity). We can say, without any fear of contradiction, that the relativistic model of geometric space–time represents one of the most important and lasting achievements of the physics of the twentieth century.

Irrespective of this convenient formal unification, however, we should never forget that space and time are two distinct physical quantities, largely different from each other. Suffice it to think, for instance, that two different bodies (or two different observers) can be instantaneously in physical contact *at the same spatial position,* even if they are characterized by *different values of the proper-time coordinate* (the opposite is impossible, instead).

We can mention, as an example of this effect, the popular relativistic result known as "the twin paradox." Two twin brothers, of identical age, first say goodbye in Rome and then meet again in Paris, after that one of them has made a long trip around the world, while the other was waiting for him without any further motion after his transfer to Paris. When they eventually meet again, the twin who traveled more is younger than the other! Therefore, they are undoubtedly in the same spatial position (Paris), but are characterized by different values of the proper time parameter (even if initially, in Rome, the proper time was the same for both twins).

This may suggest (to put it simply) that space is something "external" to the bodies, a property of Nature which exists quite independently of the individual bodies (and which allows establishing interactions among them); time, on the contrary, is an "internal" (or intrinsic) property of the bodies themselves (like charge, mass, and so on). To support this particular interpretation—time as an intrinsic property of the material bodies and of the particles composing them–there is a striking analogy with the electric charge which is worth to be pointed out.

It is well known, in fact, that a particle and its antiparticle have charges of equal magnitude but opposite sign: the electron, for instance, has a negative charge $-e$, and the antielectron (or positron) has a positive charge $+e$.

It is also well known that, in the context of the so-called Feyman diagrams (describing in a symbolic way the elementary quantum processes), particles and antiparticles are represented, for symmetry reasons, as objects propagating along the opposite space–time directions (see Fig. 4.2). On the other hand, every massive object embedded in space–time is characterized by an effective "space–time velocity"[8] whose absolute value (according to the relativistic laws) is a constant,

---

[8]We are referring to the so-called four-velocity vector (or to the associated four momentum vector) representing, geometrically, the tangent to the world-line of the given object.

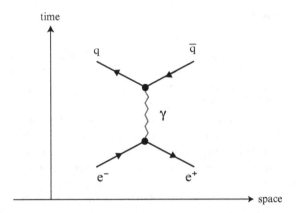

Fig. 4.2 Example of Feyman diagram: an electron (e^-) and a positron (e^+) collide and annihilate emitting a virtual photon γ which, in turn, produces a pair consisting of a new particle (q) and of its antiparticle ($\bar{q}$). The *arrows* show the direction of the effective velocity through space–time, which is positive (i.e., pointing upwards) for the particles (e^-, q), and negative for the antiparticles ($e^+, \bar{q}$)

the same for all bodies, and equal to the speed of light c. We can thus associate to a particle at rest in space a "time velocity" (characterizing the flow of proper time) equal to $+c$, and to an antiparticle (before applying the Feynman–Stueckelberg interpretation principle[9]) a time velocity $-c$.

Let us now recall that for the massless particles (like the photon) the electric charge is exactly vanishing, and the absolute value of their effective space–time velocity (i.e., of the tangent vector associated to their space–time trajectory) is also vanishing (as imposed by the laws of the relativistic dynamics). We are thus lead to conclude that, just like the charge of the elementary particles is quantized and can take the values $+e, 0, -e$ (and the corresponding multiples), also the modulus of the velocity of the temporal flow seems to be quantized, with allowed values $+c, 0, -c$.

In both cases, particles and antiparticles have values of equal magnitude and opposite sign of the considered quantities; also, in both cases, a zero value is assigned to the massless particles. A remarkable coincidence! Maybe Nature is telling us something important?

If this analogy between the electric charge and the velocity of the proper-time flow is taken seriously, and if we recall that charge can exist in Nature as an integer multiple of the elementary unit $\pm e$, then we might wonder whether in Nature there are particles (or different types of objects) whose modulus of the space–time velocity is an integer multiple of c. Or even, by recalling the existence of particles with fractional electric charges—the so-called quarks, with charges $\pm(1/3)e$ and $\pm(2/3)e$—we might wonder whether there are also the corresponding "temporal quarks," with space–time velocities of modulus equal to a fraction of c.

---

[9]According to this principle, an antiparticle of positive energy corresponds to a particle of negative energy which propagates backward in time.

In the first case the answer would seem to be negative, since it would seem to imply the existence of faster than light particles that could violate the causality principle and the basic laws of quantum-relativistic mechanics[10] (which is currently in excellent agreement with all experiments).

The answer would seem to be negative also in the second case because, if there are particles in Nature for which the limiting velocity is smaller than c, then we should also find massless particles propagating at a speed smaller than c (contrary to all experimental results so far obtained).

It should be recalled, however, that it is impossible to observe a single, fractionally charged quark in a free state, since quarks can only exist as bound states rigidly confined inside the so-called hadronic particles (for instance, protons and neutrons). Therefore, also the temporal quarks might have escaped so far a direct observation if they are strongly connected to each other and, in particular, if they are bounded in configurations of sufficiently high energy, that we have not yet been able to split up into their elementary components.

We should mention, finally, another interesting analogy concerning the possible quantization of the velocity of the flow of proper time.[11]

Quantum mechanics teach us that the spatial velocity of a particle turns out to be quantized when the particle can move only within a spatial region of finite extension, bounded by impassable physical barriers. Consider for instance a particle confined inside the impenetrable walls of a hermetically sealed container: in that case the velocity of the particle can only take values which are integer multiples of a quantity inversely proportional to the width of the container.

In the same way, also the velocity of the temporal flow (or, more precisely, the temporal component of the four-velocity vector of a particle at rest) could be quantized simply because that particle can exist only within a time interval of limited extension. Maybe (who knows?) also the life of the elementary physical systems like an electron, or a proton, is not eternal (not even in principle, as instead we currently believe), because Nature has imposed, even in time, impassable boundaries.

4.2 Maybe Space–Time Is Not Unique?

If time is an intrinsic property of all bodies, and if each particle, each test body, has its own "private" proper time, then probably the space–time itself is not a unique and absolute entity, but a notion relative to the body (or to the observer) that we are taking as reference.

In the context of relativistic physics, for instance, it is well known that the notion of space is intrinsically related to the state of motion of the observer: any

---

[10]Unless, for some reasons, "superluminal" matter with modulus of the space–time velocity equal to $2c$, $3c$, etc. does not interact at all with the "subluminal" matter present in our world, and is thus completely disconnected by our physical evidence.

[11]I wish to thank my colleague Paolo Cea for an interesting discussion on this point.

given observer identifies space (i.e., the locus of all simultaneous events) as the three-dimensional hypersurface that turns out to be orthogonal, instant by instant, to its space–time trajectory (i.e., its world-line). Different observers (or different particles), with different trajectories, have thus their own "proper" space, which corresponds to different three-dimensional sections of the four-dimensional space–time. In the presence of gravitational forces such spatial sections can also be characterized by different intrinsic geometries.

In the same way, if space–time is a four-dimensional "slice" of a higher-dimensional Universe (see Sect. 2.4), it could happen that there are particles whose world-lines are differently "bent" by the forces acting in the higher-dimensional space and that, as a consequence, "live" in space–time sections different from ours.

Is it possible, for such different and "relative" versions of space–time, to intersect and maybe interact among themselves? In the brane models suggested by string theory the answer is yes. In these models, in particular, the forces acting along the extra dimensions, from one space–time to another, are due to generalized gravitational interactions mediated by gravitons, by dilatons, and by additional strange components called "antisymmetric fields" (which have the virtue of generalizing the Abelian forces acting between point-like sources to the case of extended objects like there-dimensional branes; see in particular Sect. 6.2).

Different space–time sections may be characterized by different geometries, of course. But even in the context of the same space–time section it is possible that different particles may "feel" different geometries,[12] and thus may live in space–times which are physically different to all practical purposes.[13]

A possible example of this last effect is provided by those string models predicting a substantial contribution of the dilaton field to the total effective gravitational force. Particles with different dilaton charges will thus respond to gravity in different ways and—if they are used as geometric "probes"—they will plot different maps of the space–time, just like the test bodies embedded in different space–time geometries. In such a context we can say that to each particle (or to each class of particles with the same interaction properties) is associated a peculiar and "proper" space–time structure.

How could we interpret, geometrically, the notion of proper space–time? Just as the proper space is obtained by projecting all events on the hypersurface that is orthogonal to the (virtual) velocity flow through space–time of the given particle (or observer), in the same way the proper space–time could be obtained by projecting all events (occurring in the higher-dimensional world) on the four-dimensional space–time slice that is orthogonal to the (virtual) velocity flow of that observer through the higher-dimensional external space.

What kind of properties should characterize such a higher-dimensional external space? There are various possibilities. We could assume, for instance, that the

---

[12]Here "different" means, in particular, "not diffeomorphic," i.e., geometries which cannot be linked by coordinate transformations of differentiable and invertible type.

[13]See, e.g., the paper by N. Kaloper and K. A. Olive [36].

velocity flow in the extra dimensions is directly determined by the motion of the particle in the ordinary three-dimensional space, and then describe space–time as a four-dimensional hypersurface embedded in the so-called relativistic phase space.[14] One obtains in this way an interesting geometric scheme where, besides the existence of a limiting velocity (typical of the relativistic model of space–time), there is also a *limiting acceleration*.

However, the velocity flow in the higher-dimensional space could be completely unrelated to the ordinary motion in three dimensions: it could depend, instead, on some internal property of the particle (such as its dilaton charge). In that case, also, we would obtain a physical model where different bodies, composed of matter with different intrinsic properties, will evolve along different geodesic trajectories,[15] will follow different space–time paths, and will thus be subject to different final fates.

4.2.1 *"Relative" Singularities*

If the space–time geometry is not absolute, but relative to the considered test particle (or observer), then also important geometric properties like the presence (or the absence) of singularities, the geodesic completeness,[16] and so on, will become relative notions which are to be referred to a given observer and to the particular type of "probe" employed to explore the surrounding space–time.

The possible singularities of a given space–time geometry, in fact, are fully determined by the properties of the corresponding geodesic trajectories. If the geodesic structure of the space–time is not absolute and unique, but it is observer-dependent, then it becomes possible that the geometry turns out to be singular for a given class of observers, and regular (i.e., without singularities) for a different class of observers. The two geometries—the singular one and the non-singular one—cannot be related by general coordinate transformations (as already mentioned, they are not "diffeomorphic"): they could be connected, however, by conformal transformations.[17]

---

[14] See, e.g., the review paper by E. Caianiello [37]. Phase space is a higher-dimensional space in which the number of dimensions is doubled, since to any given coordinate is associated a new one corresponding to the so-called canonically conjugate momentum.

[15] The so-called geodesic trajectories are the world-lines determined by the space–time geometry. The set of all geodesics fully identifies a given space–time and uniquely determines its physical and geometric properties.

[16] A space–time manifold is called "geodesically complete" when all the physical geodesic trajectories can be arbitrarily extended forward and backward in time, without limits and without running into any singularity.

[17] They are transformations in which the metric is multiplied by an appropriate scalar function of the coordinates. Using this type of transformations it is possible to pass from a geometry where the trajectory of a given particle is not a geodesic to a new geometry where the same trajectory is instead a geodesic.

Such a situation may frequently occur not only in models based on string theory, but also, more generally, in all models where the gravitational interaction includes a component of scalar type, similar to the dilaton.

To give a simple example,[18] let us consider a model of Universe where the cosmic dark energy is represented by the potential energy of the dilaton, which is dominant today (see, e.g., Sect. 3.2.1). The baryonic matter, whose dilaton charge is small enough to be negligible, is coupled to the dilaton only indirectly through the gravitational energy density, and follows the geodesics of the geometry determined by the dilaton. Thus, the baryons "see" a Universe which is expanding in an accelerated way, whose curvature and a matter density are becoming smaller and smaller, and with no singularity expected to occur throughout the whole future evolution.

However, other species of matter (of non-baryonic type, possibly contributing to the dark matter density) could be directly (and strongly) coupled to the dilaton through a nonzero dilaton charge. In that case, the world-lines describing the evolution of this different matter would not correspond to the geodesics of the geometry determined by the dilaton—let us call it G—but to the geodesics of a different geometry—let us call it $\tilde{G}$—connected to G by a conformal transformation. If the dilaton charge is negative and strong enough, then one could find that the $\tilde{G}$ geometry describes a Universe expanding towards a final singular state in which the curvature is diverging!

Exotic matter, evolving along the geodesics of the $\tilde{G}$ geometry, would thus be doomed to be absorbed by the future singularity, while ordinary matter, following the geodesics of the G geometry, would remain unscathed. If we are made of "good" matter, without singularities in its own future and designed to reach the "heaven" of an infinite time evolution, we might see[19] sooner or later the explosion of "bad" matter, doomed to fall into the "hell" of the future singularity and to disappear forever.

Let us hope that it is not the opposite case!

---

[18]This example was discussed in a paper I wrote in 2004 [38].

[19]Not us, but our descendants! Because the possible future singularity would be distant in time billions of years.

Chapter 5
Strings and Fundamental Interactions

Why do we find in Nature only four different types of fundamental interactions (corresponding to gravitational, electromagnetic, strong, and weak nuclear forces)? And why do these forces behave just in the way we know, namely why does the electric field obey Coulomb's law, the gravitational field Newton's law, and so on for the other interactions?

String theory is currently the only theoretical framework able to answer questions of this kind.

According to string theory, in fact, all particles present in Nature—and, in particular, the photon that mediates electromagnetic interactions, the graviton that mediates gravitational interactions, and so on—must correspond to possible physical states obtained by quantizing the oscillations of an elementary string. It is thus the quantum version of a string model that "decides" what particles are allowed and what are not (and thus, also, what types of interactions can exist in Nature).

If we consider the "spectrum" (i.e., the set of all possible states) of a quantized open string we find, in particular, a state describing a particle of zero mass and spin 1 which exactly corresponds to the photon, carrier of the electromagnetic interactions; in the spectrum of a closed string we find state which describes a particle of zero mass and spin 2, which can be interpreted as the graviton; and so on. In the string spectrum, however, there are no states describing (for instance) massless particles of spin 3: and there is indeed no evidence, in Nature, of the presence of the corresponding long-range force (which would be characterized by rather unusual physical properties).

But why do the existing forces precisely follow those dynamic laws that we are observing, and not other laws? More explicitly, why—for instance—the electromagnetic field obeys Maxwell's equations and the gravitational field obeys Einstein's equations? These equations have been introduced (and are still used) because they are in good agreement with experiments; however, they are not the only equations eligible, in principle, to describe phenomena of gravitational or electromagnetic type.

Even in this respect string theory is surprising because, besides telling us what types of force fields can (and must) exist, it is also able to tell us which laws

M. Gasperini, *Gravity, Strings and Particles*, DOI 10.1007/978-3-319-00599-7_5,
© Springer International Publishing Switzerland 2014

(i.e., which equations) must be satisfied by those fields. How can it do this? Using, again, quantum theory and some peculiar physical properties of the extended objects that will be illustrated in the following sections.[1]

5.1   How to Quantize Extended Objects

Strings—either open (as small wires with disjoint endpoints) or closed (as small elastic rings)—are elementary objects extended along one spatial dimension (see Fig. 5.1). Like every extended object they correspond, from a dynamic point of view, to "constrained" systems: hence, their quantum description is always more elaborate than required for point-like objects (like particles) or fields.

Each point of a string, in fact, describes with its time evolution a single world-line. The set of all world-lines spans the so-called "world-sheet" of the string, a two-dimensional surface which can be open, or closed onto itself like a cylinder, and which represents the overall trajectory of the string through space–time (see Fig. 5.2).

For a complete (and mathematically consistent) description of the string motion we must then meet three requirements: (1) provide the position of all points of the string as a function of time; (2) specify the behavior of the string endpoints (they can be fixed or free to move); (3) take into account the shape (or better the geometry) of the world-sheet surface.

The first requirement can be satisfied by writing down the usual Euler–Lagrange equations, similar in type to those describing the motion of a particle or the time-evolution of a (classical) field of force.

The second requirement is met by specifying the so-called boundary conditions (already introduced in Sect. 2.4). For closed strings such conditions impose on the coordinate spanning the spatial dimension along the string extension to be "periodic", i.e., to return to the same value after each round. For open strings, instead, they impose on the string endpoints to stay at fixed positions (Dirichlet conditions), or (if they are not fixed) to move in such a way that there is no flow of kinetic energy out of the string through its ends (Neumann conditions).

Finally, the third requirement leads us to a series of conditions—called "Virasoro constraints" [43]—imposing on the points of the string to move "coherently", to avoid breaking the space–time fabric of the string world-sheet. In fact, if the different points of the string were moving independently one from the other, in a completely uncorrelated way, then the string evolution would not necessarily reproduce a compact two-dimensional surface: in that case, we might not regard the

---

[1] We will try to be as accurate and as detailed as possible, but we will limit here to a very qualitative introduction to the existing models of strings and superstrings. The readers interested in a deeper study of the physical and mathematical aspects of such models are referred to the textbooks [39–42] listed in the references.

Fig. 5.1 Example of open
string and closed string

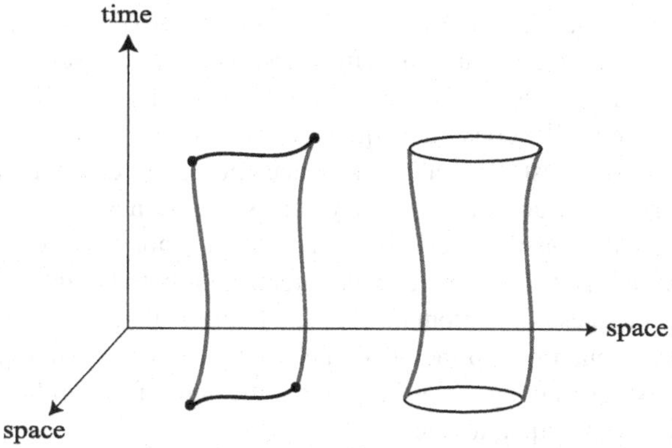

open string   closed string

Fig. 5.2 Examples of
possible world-sheets for an
open string (*left*) and a closed
string (*right*)

string as an elementary extended object, but only as a collection of many elementary point-like objects.

The Virasoro constraints, as we shall see, will play a fundamental role also in the process of string quantization. Before considering quantization, however, it may be useful to recall that the general solution of the classical equations of motion, taking into account the appropriate boundary conditions, can be written (for both open and closed strings) as a discrete sum of infinitely many terms.

Each term corresponds to a possible oscillation mode of the string, a tiny "ripple" of constant wavelength and amplitude which propagates from one endpoint to the other in the case of open strings, or in circle (in both senses) in the case of closed strings. The frequencies of such waves take discrete and equally spaced values, varying from zero to infinity. The sum of all waves, weighted by appropriate amplitudes, fully determines the particular vibrational state that the string is currently experiencing.

When applied to this general solution the Virasoro constraints provide a (discrete) series of infinitely many conditions, and each condition, in turn, can be written as a discrete sum of infinitely many contributions (see, in particular, the reference books [39–42]). Such conditions relate among themselves the amplitudes of the different oscillation modes, ensuring that all vibrations stay confined on the string world-sheet. The first of these conditions, in particular, is closely connected to the total energy of the string and relates the mass of the string to its total relativistic momentum.

In order to provide an appropriate description of the string motion in a quantum context we may follow the so-called canonical procedure, and replace all classical string variables (position, momentum, etc.) with suitable quantum operators defined in the corresponding Hilbert space.[2]

By applying the relevant commutation rules[3] we then find that a string, from a quantum point of view, is equivalent to a system of infinitely many elementary harmonic oscillators, each one having its own frequency and its own spectrum of discrete energy levels. The generic state of a quantized string can thus be represented as a state including an arbitrary number of (possibly excited) oscillators, each of them arranged in a well-defined energy level. The set of all possible combinations of oscillators represents the so-called Fock space, which is a particular type of Hilbert space for the considered quantum system.

The most general Fock space, however, could include states which are not physically acceptable as they correspond to negative probabilities (they are the so-called "ghost" states). If such states are not removed, we would obtain a model which is not only hard to be interpreted, but also in contrast with the fundamental principles of quantum mechanics. In particular, the model would be in contrast with the "unitariety" principle that characterizes the evolution of a quantum system, and ensures (among other things) that the sum of the probabilities of all possible states is always equal to one.

It is precisely at this point that, providentially, the Virasoro constraints come into action.

The quantum states of the string, in fact, must satisfy the Virasoro constraints precisely as the classical solutions of the string equations of motion. In the quantum version of the string model the Virasoro constraints are represented by appropriate operators, which, acting on the states of the Fock space, decide which states are admissible for the string and which are not. The allowed states are those describing oscillations along spatial directions[4]: this automatically eliminates all possible "ghost" states, associated with oscillations along the temporal direction.

The Virasoro operators (namely, the operators representing the Virasoro constraints) also determine the allowed energy levels of the string spectrum, and fix the values of the masses associated with the physical quantum states. This implies, in particular, that they control the masses of the particles represented by the various states of the quantized string model.

Regarding this point let us recall that, in order to switch from the classical to the quantum representation of the Virasoro constraints, we need to "regularize" the quantum Virasoro operators, by subtracting all possible terms that give an infinite contribution.

---

[2]It is the set whose elements are the possible quantum states of the considered physical system.

[3]The rules telling us how the action of a product of operators changes if we reverse the order of the factors.

[4]Using the invariance under general coordinate transformations we can further eliminate the "longitudinal" oscillations, and we are eventually left only with "transverse" oscillations, i.e., oscillations along the spatial directions orthogonal to the string itself.

Considering for instance the operator corresponding to the first term of the series of Virasoro constraints—i.e., the one which directly represents the total energy of the string—we find the contribution of the so-called zero-point energy[5] of all (infinitely many) quantum oscillators associated to the string. Assuming that the string is embedded in a globally flat space–time, not deformed by the presence of gravity, such an infinite energy contribution can be appropriately subtracted by applying the formal methods of ordinary quantum field theory.

In that case, the constraint provided by the regularized Virasoro operator allows us to express the mass of the string states as a function of three important parameters:

- the quantum number N (a non-negative, integer number) determined by the energy levels of the oscillators possibly present in the given state;
- the string tension (i.e., the energy per unit length of the string), a constant quantity which has the same value for all the elementary strings of the considered model;
- the total number D of dimensions of the space–time in which the string is embedded.

The most surprising aspect of the mass spectrum obtained in this way is that the number of space–time dimensions D *cannot* be arbitrarily prescribed, but is rigidly fixed by the type of string model we are considering. This means, more explicitly, that a string "chooses" by itself the number of dimensions in which to live! The string is not satisfied with a space–time with a different number of dimensions: if the preferred value of D is not respected, the corresponding model is no longer consistent with the basic principles of quantum mechanics and relativistic dynamics.

This is a revolutionary result, as we might have expected—in view of many other important examples—that a fundamental physical systems can always be described in an arbitrary number of space–time dimensions. For a string, instead, it is not so. How do we reach such a conclusion?

Let us take an explicit example considering the so-called bosonic string model, a model where the string system is fully described by variables of bosonic type (such as the space–time coordinates of the various points of the string). Let us focus on the particular case of open strings and consider the first excited level of their quantum spectrum (namely, the state characterized by the quantum number $N = 1$).

Such a state is represented by a single oscillation mode and, from the point of view of a transformation of the coordinates of the D-dimensional space–time in which the string is embedded, the state behaves as vector. The Virasoro constraints imply, in particular, that this vector has to be of "transverse" type, i.e., orthogonal to the direction of the string motion through the external space–time.

As already stressed, however, any given state can always be represented in terms of one-dimensional oscillators vibrating along the spatial dimensions orthogonal to the string itself. Such dimensions (subtracting from the total number D the temporal direction and the "longitudinal" spatial direction, oriented along the string

---

[5]It is the same energy already mentioned at the beginning of Sect. 3.3.

extension) are in total $D - 2$: hence, the vector representing the considered string state actually describes $D - 2$ physical degrees of freedom, and thus contains only $D - 2$ independent components (instead of the D components typical of a vector in D-dimensional space–time).

But—according to the laws of the relativistic dynamics—a vector state which on one hand satisfies the transversality condition and, on the other hand, has two independent components less than what expected from the total number of dimensions, must necessarily represent a physical system which is massless and propagates at the speed of light. A vector state of this kind, incidentally, can be exactly identified with the state representing a photon, the quantum of electromagnetic interactions.

If we look now at the mass spectrum determined by the Virasoro constraints,[6] we find that the state with $N = 1$ can be massless only if $D = 26$! The model of the open bosonic string is thus compatible with quantum mechanics and relativistic dynamics only if formulated in a space–time with 26 dimensions.

The same result is also reached considering bosonic strings of closed type. The first excited level contains, besides a scalar field (the dilaton) and an antisymmetric field (the so-called axion), an additional, symmetric tensor field which is "transverse", and which turns out to be massless—as required by the relativistic symmetries—only if $D = 26$, as in the previous case. Such a tensor field, by the way, satisfies all physical properties required to represent the graviton, the carrier of the gravitational interactions.

Hence, the model of (closed or open) bosonic string comes out without "ghost" states (i.e., negative probability states) thanks to the Virasoro constraints; in addition, the excited levels of this model are consistent with the laws of quantum and relativistic mechanics that provided the external space–time has 26 dimensions. As a byproduct, the model includes in its spectrum the states needed to represent electromagnetic and gravitational interactions.

Despite these achievements, unfortunately, the model has an important shortcoming: the fundamental energy-level of the bosonic string, i.e., the state with quantum number $N = 0$, is associated with a mass whose square takes a negative value. It follows that the mass of the ground state turns out to be imaginary, for both open and closed strings of bosonic type.

States with imaginary mass (called "tachyons") are problematic, not only from the point of view of relativistic theory (as they can propagate faster than the speed of light), but also from the point of view of quantum theory (as the amplitude of these states can grow exponentially in time, making the model unstable). They are to be removed, just like the ghost states, if we want a model proving to be physically acceptable and consistent with current theoretical paradigms.

---

[6]The reader interested in the exact expression of the mass spectrum for a bosonic string, and on its explicit dependence on D and N, is referred to the books [39–42] listed in the final bibliography.

The suppression of all tachyonic states is automatically implemented in the context of the so-called superstring models (i.e., models of supersymmetric strings) that will be introduced in the following sections.

5.2   Supersymmetry and Higher-Dimensional Spaces

The fundamental level of the bosonic string is characterized by a negative eigenvalue of the squared-mass operator, and thus corresponds to a tachyon, namely to a state of imaginary mass. The next energy level (i.e., the first excited level with $N = 1$) corresponds to a massless state, while all subsequent energy levels ($N = 2, 3, \ldots$) correspond to states characterized by a real and positive value of the mass parameter.

The tachyon problem is thus confined to the negative eigenvalue obtained in the lowest energy level of the spectrum. Recalling that the state of minimal energy of a supersymmetric system is typically characterized by a zero eigenvalue (see Sect. 3.3), we could think of eliminating tachyons by implementing a generalized version of the string model which includes supersymmetry.

To this purpose, let us first notice that the dynamical description of a string requires two types of bosonic variables: the coordinates, giving the positions of the various points of the string in the external space–time, and the metric, describing the geometry of the string world-sheet. A supersymmetric version of the string model must thus contain appropriate fermionic partners for each coordinate, and further (distinct) fermionic partners for each component of the world-sheet metric.

All these new variables must behave as spinor fields (namely, as fields describing particles of half-integer spin) under the coordinate transformations performed on the string world-sheet. In addition, the partners of the coordinates must transform as vector-like objects in the external space–time, while the partners of the world-sheet metric as tensor-like objects in the string world-sheet.

The inclusion of such new ingredients leads us to a geometric structure extremely rich in symmetries. We obtain, in fact, a model which is invariant under coordinate transformations in the string world-sheet and in the external space in which the string is embedded; invariant under the exchange of bosonic and fermionic variables; invariant under local "rescaling" transformations (i.e., local dilatations or contractions) of the world-sheet metric and of its fermionic partners. This latter symmetry is also called "superconformal symmetry."

A physical system which enjoys all these symmetries is called "superstring" and—as we shall see—turns out to be compatible with the laws of quantum and relativistic dynamics only if embedded in a space–time with $D = 10$ dimensions. And, perhaps even more interesting, it turns out that there are only *five different types* of consistent superstring models. Let us try to explain why and how we come to identify these five types.

The time evolution of a superstring, just like the evolution of a string, has to be determined by imposing appropriate constraints and boundary conditions. For the constraints, in particular, we find that the Virasoro constraints of the bosonic string are to be promoted to "superconstraints" involving both the bosonic and fermionic variables, which get mixed among themselves. The boundary conditions, instead, must be separately applied to the two types of variables.

Let us first consider the boundary conditions. For the bosonic variables they are identical to those discussed in the previous section. For the fermionic variables we must impose additional boundary conditions, which can be satisfied in various ways.[7]

- In the case of open superstrings there are two choices: we can impose on the fermionic variables to be either periodic or antiperiodic, i.e., we can impose that the fermionic variables, at each end of the string, take either the same value or the opposite value. In the first case we say that we are applying Ramond (R) boundary conditions, in the second case Neveau–Schwarz (NS) boundary conditions.

- In the case of closed superstrings there are instead four possible choices, since the periodicity (R) or anti-periodicity (NS) condition has to be separately imposed on the vibration modes propagating along the string in clockwise and in counterclockwise direction (such oscillations are also called, conventionally, "right-moving" and "left-moving" modes). We can thus impose boundary conditions which are both of periodic type (R–R), or both of antiperiodic type (NS–NS), or of mixed type (either R–NS or NS–R).

We have then to take into account the Virasoro constraints. In the quantum version of the superstring model they are represented by operators that must be regularized, in order to subtract all infinite contributions, by applying the standard commutation rules for the bosonic variables and the appropriate anti-commutation rules for the fermionic ones. The action of these operators, as in the case of the bosonic string, removes the ghost states and determines the allowed energy levels of the quantized superstring.

The Virasoro operators can be conveniently separated into a bosonic and a fermionic part. The bosonic part is identical to the one of the bosonic strings considered in the previous section. The new fermionic part differs from the bosonic one in two important respects: the first is that, when expanded into a sum of infinitely many terms, it involves discrete sums over half-integer numbers $(1/2, 3/2, 5/2, \ldots)$; the second is that it explicitly depends on the assumed boundary conditions.

Imposing the quantum Virasoro constraints, and including the whole set of bosonic and fermionic components, we find that the energy spectrum of the superstring depends in general by various parameters. In addition to the string tension and to the number D of space–time dimensions, in fact, the spectrum depends on the (non-negative) integer number N that specifies the energy levels of the bosonic oscillations, on the (non-negative) integer number N_R that specifies the levels of the fermionic oscillations with periodic (R) boundary conditions, and on

---

[7] As shown in the works of Ramond [44], and of Neveau and Schwarz [45].

the (non-negative) half-integer number N_{NS} that specifies the levels of the fermionic oscillations with antiperiodic (NS) boundary conditions.

The new form of the energy spectrum immediately leads us to some important physical results. First of all, proceeding as in the bosonic case,[8] we can conclude that the allowed states of the quantum superstring are compatible with the laws of relativistic mechanics only in a space–time with $D = 10$ dimensions.

In addition, in view of the dependence of the energy spectrum on the boundary conditions, we can conclude that there are two different types of spectrum for the open superstring (depending on if we are using R or NS conditions), and four different types of spectrum for the closed superstring (depending on if we are using R–R, or NS–NS, or R–NS, or NS–R conditions). And it is worth stressing that such different spectra include states of different statistical type. In particular:

- In the case of the open superstring the states corresponding to NS-type levels are represented by fields transforming as tensors in the external space–time in which the string is embedded, and are thus states with statistical properties of bosonic type. The states corresponding to R-type levels are represented, instead, by fields transforming as spinors in the external space–time, and are thus states with statistical properties of fermionic type.
- In the case of the closed superstring the states corresponding to levels of R–R and NS–NS type are represented by tensor fields and are thus states with statistical properties of bosonic type, while the states corresponding to levels of R–NS and NS–R type are represented by spinor fields, and are thus states with statistical properties of fermionic type.

We should emphasize, at this point, that the presence of states describing spinor objects in the external ten-dimensional space–time is of crucial importance for a possible interpretation of the quantum superstring model as a unified model of all fields and all physical interactions.

The fundamental components of microscopic matter (the so-called quarks and leptons) are indeed represented by spinor fields. The superstring model, though formulated in terms of classical variables transforming as vector in the external space–time and as spinors in the intrinsic string world-sheet, also *automatically* includes—once quantized—states transforming as *spinors* in the *external space–time*. It is thus a model able to describe within a single theoretical scheme not only the fundamental forces of Nature (represented by vector or tensor fields in the space–time) but also their elementary quantum sources (represented by spinor fields in the space–time). A virtue that is unique among all tentative models of unified theories so far proposed.

However, in spite of the mentioned virtues, the above model of supersymmetric string is still unsatisfactory for various reasons.

---

[8]Namely, considering the levels associated with states of vector or tensor type, noting that those states are "transverse," that the number of independent vector components is $D - 2$, and imposing on those states to be massless.

First of all, considering all possible spectra (of open or closed strings, of R-type or NS-type, etc.), we find that some of them are characterized by a massless fundamental level, but other spectra are still characterized by a fundamental level with a negative squared-mass eigenvalue: hence, tachyons are not completely removed. Second, there are a few statistical inconsistencies, due to the presence of bosonic states connected to each other through the action of an odd number of fermionic operators.

Third, if we compare the number of degrees of freedom (i.e., of independent components) of the bosonic states—associated to spectra of NS type—with the number of fermionic degrees of freedom—associated to spectra of R type—we find that the two numbers are different, in general, even inside the same energy level. This prevents the possible presence of configurations which are supersymmetric in the external ten-dimensional space–time, despite the fact that the model is (in all respects) supersymmetric in the two-dimensional world-sheet of the string.

To get rid of the above difficulties we can apply a procedure, called "GSO projection,"[9] that provides an additional "filter" to select the physically allowed states of the quantum superstring.

The effects of this procedure are twofold: on one hand, the GSO projection removes from the spectra of NS type all states obtained by applying to the fundamental state an odd number of fermionic operators; on the other hand, it removes from the spectra of R type half of their fermionic components, leaving to each spinor field only one of its two "chiral" components[10] (the positive one or the negative one).

The above prescriptions, which might seem arbitrary and unmotivated if imposed a posteriori on the spectrum, actually emerge from the outset (and are automatically satisfied) if the model is formulated so as to be supersymmetric on the whole space–time, and not only on the string world-sheet.

In any case, the GSP "treatment" has a miraculous effect: removes all tachyons, setting to zero the mass of the fundamental level for any type of spectrum; fixes the statistical anomalies mentioned above; finally, equalizes the number of bosonic and fermionic degrees of freedom inside each level of the spectrum, thus making the system suitable to the presence of space–time supersymmetry (by the way, it is just because of that symmetry that tachyons are completely removed from the spectrum).

---

[9]The acronym GSO is due to the names of the authors, Gliozzi et al. [46].

[10]Each spinor field can be always decomposed in two chiral components, called "right-handed" and "left-handed" (or "positive" and "negative"). These components represent opposite physical configurations in which the intrinsic angular momentum of the spinor is, respectively, parallel or antiparallel to the direction of motion. The description of a massive spinor field requires the inclusion of both chirality components. The spectrum of the superstring, however,contains massive fermions even after the selection produced by the GSO rule: they are obtained by combining together different spinors of opposite chirality, which are always present inside the massive energy levels of the spectrum.

5.3   The Five Superstrings

We may wonder, at this point, whether the boundary conditions, the Virasoro constraints, and the GSO projection are enough (taken together) to uniquely define the string model, or whether, instead, there is still some arbitrary choice to be made.

5.3.1   Type IIA and Type IIB Superstrings

Let us start with closed superstrings, considering in particular those states which contain spinor fields. When applying the GSO projection, which removes half of the fermionic components, we can still choose whether to leave in the same level spinors of *opposite* chirality or of *the same* chirality. In the first case we obtain a model called type IIA superstring, in the second case a model called type IIB superstring. What are the differences?

The differences already emerge at the level of the ground states of these two strings if we take into account the physical fields (i.e., the particles) appearing in the bosonic and fermionic sectors of the spectrum.

Consider in fact the bosonic sectors, which—as previously noticed—can be of R–R type or NS–NS type. The ground state of the NS–NS spectrum is the same for the two types of superstrings, and describes a multiplet of three fields: a scalar (the dilaton), a symmetric tensor (the graviton), and an antisymmetric tensor[11] of rank 2 (the so-called Kalb-Ramond axion). The content of the R–R spectrum, instead, depends on the type of superstring: the ground state of the type IIA string describes a vector field and an antisymmetric tensor of rank 3; the ground state of the type IIB string describes a scalar field, an antisymmetric tensor of rank 2 and another antisymmetric tensor of rank 4. The number of degrees of freedom (i.e., the number of independent components) is obviously the same in both cases.

Consider now the fermionic sectors of the spectrum, which can be of R–NS type or of NS–R type. The ground state of the R–NS spectrum is the same for the two types of superstrings and contains two spinor fields of the same chirality (for instance, positive chirality): one is the "dilatino" (a particle of spin $1/2$, representing the supersymmetric partner of the dilaton), and the other one is the gravitino (a particle of spin $3/2$, the supersymmetric partner of the graviton). The ground state of the NS–R spectrum, instead, depends on the type of superstring: it still contains a dilatino and a gravitino of the same chirality, but such a chirality is the

---

[11]An antisymmetric tensor field of rank n is a bosonic object represented by a variable with n indices which behaves as a tensor under the coordinate transformations of the ten-dimensional space–time, and which is characterized by the following property: it changes sign whenever any two of its indices are exchanged between themselves. In the case of an antisymmetric tensor of rank 3, for instance, we have: $A_{\alpha\beta\gamma} = -A_{\beta\alpha\gamma} = -A_{\gamma\beta\alpha} = -A_{\alpha\gamma\beta}$.

opposite to that of the NS–R spectrum (i.e., negative) for the type IIA, and the same as that of the NS–R spectrum (i.e., positive) for the type IIB superstrings.

Putting together the physical content of the various sectors of the spectrum, we can summarize the differences by saying that type IIA superstrings describe, with their ground state, the dilaton, the graviton, the axion, a vector field, an antisymmetric field of rank 3, two dilatinos of opposite chirality, and two gravitinos of opposite chirality. Type IIB superstrings describe, instead, the dilaton, the graviton, the axion, a scalar field, two antisymmetric fields of rank 2 and 4, two dilatinos of the same chirality, and two gravitinos of the same chirality.

So, the above types of closed superstrings certainly include the fields needed to describe gravity (and the description is done within an elegant supersymmetric formalism). But what about the fields needed to describe the other fundamental interactions? The so-called Abelian and non-Abelian gauge fields (see Sect. 2.3.2) seem to be absent. It is precisely in this respect that—fortunately enough—open superstrings come to our rescue.

5.3.2 Type I Superstring

An open superstring, in fact, can carry point-like "charges" (of electric type, nuclear type, and so on) which are located on the ends of the string, and which are sources of gauge fields associated to appropriate symmetry groups. The allowed symmetry groups, and the corresponding types of interactions, depend on a property of the string called "orientation."

Let us recall, in this regard, that a model of (open or closed) string is said to be "non-oriented" if it is invariant under the inversion of the coordinate that specifies the spatial positions of the various points of the string on the world-sheet surface. As a consequence, the spectrum of a quantized non-oriented string contains only those states that are invariant under such transformation, while the spectrum of an oriented string contains also non-invariant states.

Using open strings, either oriented or non-oriented, we can describe different types of gauge interactions associated, in general, with non-Abelian gauge groups. A model which is complete and consistent also in a quantum context, however, cannot contain only open strings, as the two ends of a string could spontaneously join themselves and form a closed string. Can we put together open and closed strings in a model which is both supersymmetric and quantistically consistent? The answer depends on the orientation of the strings that we are considering.

In particular, we cannot combine oriented strings of open and closed type, as they correspond to different supersymmetric frameworks: the spectrum of open superstrings contains indeed a single gravitino, while the one of closed superstrings (illustrated in the previous section) contains two gravitinos . In order to include open and closed superstrings in the same model we have thus to "cut" the spectrum of closed superstrings, eliminating, in particular, the surplus of gravitinos. Such a procedure, once correctly performed according to the rules of quantum mechanics and supersymmetry, is precisely equivalent to imposing on the closed string to be non-oriented.

For a consistent model, open superstrings are thus to be paired with *closed non-oriented* superstrings, i.e., with closed strings which are invariant under reflections of the spatial coordinate on the string world-sheet. Such an invariance property can be satisfied by type IIB superstrings (whose spectrum contains spinors of the same chirality), and not by type IIA superstrings (whose spectrum contains spinors of opposite chirality). Hence, we must combine open superstrings with closed, non-oriented superstrings of type IIB.

Non-oriented type IIB superstrings, on the other hand, are affected by formal problems due to the presence of infinities and quantum anomalies.[12] Such problems can be eliminated only with the coupling to open superstrings bearing charges of the non-Abelian gauge field associated to the $SO(32)$ symmetry group[13] (see e.g., the textbooks [39–42] for a detailed discussion). But such a particular gauge, group turns out to be compatible with open superstrings only if they too are of non-oriented type!

Summarizing the above results, we can say that it is possible to formulate a supersymmetric model, fully consistent from the quantum mechanic point of view, putting together open strings and type IIB closed strings, both of non-oriented type. The resulting model provides—in ten space–time dimensions—a unified description of the gravitational interactions and of the interactions mediated by the gauge field of the non-Abelian $SO(32)$ group. Such a model is called "type I superstring" (see the illustrative diagram of Fig. 5.3).

If we look at the energy spectrum of this type of superstring we find, in the (massless) ground state level, the following fields: the dilaton, the graviton, a single antisymmetric tensor of rank 2, a dilatino and a gravitino of the same chirality, and, finally, a vector gauge field for the $SO(32)$ group with the associated supersymmetric partner, the so-called gaugino (of spin $1/2$).

It is important to stress that all these fields are objects defined in a ten-dimensional space–time, hence the model has to be "dimensionally reduced" from 10 to 4 to be applied to our four-dimensional physical Universe, especially if we want to compare its phenomenological predictions with presently available observations.

To this purpose we could assume, for instance, that six spatial dimensions are "rolled up" and compactified on very small length scales, and that only the remaining four dimensions turn out to be accessible to current observations (see also Sect. 2.3). If the compactified six-dimensional space is characterized by a suitable geometric structure, we find that the ten-dimensional gauge field of the

---

[12]We say that a model is affected by a "quantum anomaly" when a symmetry, which is present in the classical version of the model, is violated by the quantization procedure and thus disappears in the quantized version of the model.

[13]It is the group representing all possible rotations that can be performed within a Euclidean space with 32 dimensions.

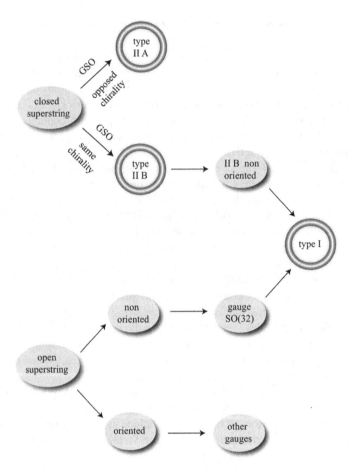

Fig. 5.3 A simplified "family tree" for type I superstring, a type of string resulting from the combination of two particular subspecies of closed and open strings

$SO(32)$ group can be decomposed into various lower-dimensional fields, including, in particular, the four-dimensional gauge fields needed to describe the familiar electromagnetic, weak, and strong interactions. In that case, type I superstring could represent a promising model for a unified description of all interactions.

However, type I superstring is not the only string model able to include non-Abelian gauge fields into its spectrum.

5.3.3 The Two "Heterotic" Superstrings

With the cases so far discussed, it might seem that we have exhausted all possible versions of consistent superstring models. We have found, in particular, that a consistent model of closed superstring has to be of type IIA or type IIB; if we want to include open superstrings they must be non-oriented, and they are always to be associated with closed, non-oriented type IIB superstrings.

However, there is a further possibility to obtain a consistent supersymmetric model even including *oriented* closed strings: the so-called model of "heterotic" superstring.[14]

Let us recall, to this purpose, that the closed bosonic string has oscillation modes propagating along the string in two opposite directions (clockwise and counterclockwise), that we shall conventionally distinguish with the names of "right-moving" and "left-moving" modes (as in Sect. 5.2). Let us then try to implement supersymmetry only for one half of these oscillations, introducing the appropriate fermionic partners—for instance—for the right-moving modes only, leaving the left-moving modes devoid of the analogous supersymmetric completion.

As we know, a consistent quantization of the bosonic oscillations needs 26 space–time dimensions. Supersymmetric oscillations, instead, need ten space–time dimensions: hence, for consistency, the set of supersymmetric modes will oscillate within a nine-dimensional subspace of the whole space–time manifold.

The additional 16 spatial dimensions are at the disposal of the left-handed bosonic oscillations only. We can assume that these extra dimensions are characterized by a compactified geometry but, recalling the discussion of Sect. 2.3.2, we must impose on such geometry the condition of zero Ricci curvature, to ensure that the ten-dimensional space–time—where the supersymmetric part of the string is living—stays flat and infinitely extended.

The bosonic, left-moving oscillations propagating through the 16 compact dimensions are to be quantized by imposing appropriate boundary conditions. And here we have two possibilities, due to the fact that the oscillations of such bosonic modes are equivalent (from the point of view of the quantum effects on the string world-sheet) to the oscillations of 32 real spinor fields of fixed chirality.[15]

If we impose on all 32 spinors *the same* boundary conditions (either periodic or antiperiodic), then we obviously obtain a model which is invariant under the action of the $SO(32)$ group, exchanging the spinors one with another. But there is also another, possible type of boundary condition which can be consistently imposed: we can divide the spinors into two groups of 16 elements, and impose periodic conditions on one group, and antiperiodic conditions on the other. The resulting configuration is then invariant under the transformations of the so-called $E_8 \times E_8$ group.[16] No other consistent condition is possible.

In the first case we obtain a type of heterotic superstring which includes the gauge field of the $SO(32)$ group. In the second case we obtain another type of

---

[14]The name, derived from the ancient Greek language, means that we are combining things which are different and seemingly incompatible.

[15]Fields of this type are also called "Weyl–Majorana spinors."

[16]The E_n group is the group representing all possible translations and rotations that can take place in an n-dimensional Euclidean space.

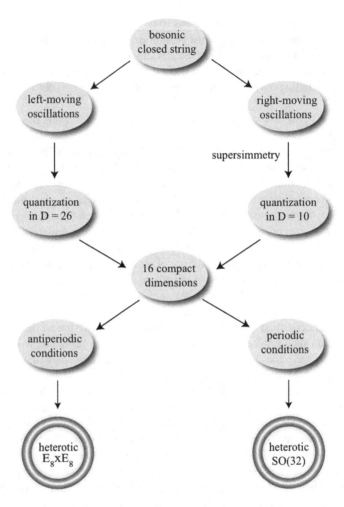

Fig. 5.4 The "family tree" of the two possible types of heterotic superstrings that include gauge fields associated to the symmetry group $SO(32)$ or $E_8 \times E_8$. The two types of string have the same "ancestors", but differ for the boundary conditions imposed on the fields living in the 16 compact dimensions

heterotic superstring, which includes the gauge field of the $E_8 \times E_8$ group. The close relationship between the two models is summarized and illustrated by the diagram of Fig. 5.4.

What are the physical differences between these two types of heterotic super-strings and the other types of superstrings so far considered?

If we look at the ground state of the heterotic superstrings (corresponding, again, to the massless, lowest energy level), we find that it describes the following fields: the dilaton, the graviton, an antisymmetric tensor field of rank 2, the dilatino, the gravitino, the gauge vector field, and the gaugino. The physical content is thus exactly the same as that of the type I superstring, with the only difference that the gauge group may correspond either to $SO(32)$ or to $E_8 \times E_8$.

However, if we take into account the mutual interactions of these fields, the physical scenario is different, as the coupling of the dilaton to the gauge field and to the antisymmetric tensor is different in the two cases. Nevertheless, it is important to stress that the physical configurations described by the type I superstring and the heterotic superstring are related—at least for the ground state—by a suitable "conformal" transformation of the geometry and by a simultaneous change of sign of the scalar dilaton field.

This change of sign, in particular, is equivalent to the inversion of the effective coupling constant that controls the intensity of all string interactions[17]: thanks to this inversion, the string switches from a large to a small value of the effective coupling (or viceversa), and thus from the strong-coupling to the weak-coupling regime (or viceversa). This suggests that type Ia and heterotic superstrings are complementary, in the sense that the physical phenomena described by one type of string in the regime of strong interactions are described by the other type of string in the opposite regime of weak interactions (the so-called perturbative regime), and viceversa.[18]

We should finally notice that also the heterotic superstring, just like the type I superstring, can be used as a possible starting point towards a unified description of all interactions. A realistic model requires of course the further compactification of six dimensions, needed to get back from the total number of ten to a final (more familiar) number of four space–time dimensions. In this regard, the heterotic superstring with the gauge group $E_8 \times E_8$, and with 6 spatial dimensions compactified in such a way as to form a space of Calabi–Yau type,[19] seems to be the model with the best chances of reproducing the low-energy phenomenology of the fundamental interactions that we are currently observing.

5.4 Conformal Invariance and Equations of Motion

From what we have seen so far, it seems that the superstring models are able (at least in principle) to fulfill the old Einstein's dream of describing in a unified and geometric way gravity, electromagnetism, as well any other possible type of fundamental interaction.

Let us recall, in fact, that the superstring models are characterized by the following properties.

---

[17]As we shall see in Sect. 5.4.1, such a coupling constant is proportional to the exponential of the dilaton field, and is thus inverted if the sign of the dilaton is changed.

[18]The two types of strings are actually related by a so-called duality transformation (see Sect. 5.5).

[19]It is a compact space which has vanishing Ricci curvature, and which can be described by a complex geometric structure parametrized by three real and three imaginary coordinates.

- Like all string models, they *unavoidably* include in their spectrum a symmetric tensor field (the graviton), carrier of the gravitational forces.
- Being supersymmetric (to avoid tachyons), they *unavoidably* include also the spinor fields needed to describe the fundamental fermionic components of matter (quarks and leptons).
- Finally, for a consistent (i.e., ghost free and anomaly free) quantization, they necessarily select a *unique* set of allowed gauge symmetries: those associated with the non-Abelian $SO(32)$ and $E_8 \times E_8$ groups, in a space–time with $D = 10$ dimensions. The dimensional reduction, with the appropriate compactification of six spatial dimensions, leads then to reduced symmetry configurations including those of the $SU_3 \times SU_2 \times U_1$ group,[20] i.e., of the group describing just the electroweak and strong interactions that we are experiencing in four dimensions.

Besides predicting the existence of known particles (sources and carriers of forces that we have already experienced), described by the lowest energy levels of the string spectrum, the superstring models also arouse our curiosity and stimulate our imagination with the infinite number of physical states associated to the full string spectrum. This infinite series of states describe particles of growing masses and growing spins[21] that we have not yet discovered (and that perhaps we will never discover until this requires energy scales too high to be experimentally accessible).

But the most impressive aspect of string theory—and let me tell this from the point of view of a physicist used to everyday working with the equations of (classical and quantum) field theory—is not so much to predict which particles, or fields, can exist in Nature, but rather to fix in a unique and complete way the equations of motion of all fields (bosons and fermions, massless and massive) present in the string spectrum.

In fact, the superstring models not only tell us that we should find in Nature gravitational fields, spinor fields, Abelian and non-Abelian gauge fields; they also tell us that a string cannot interact with those fields without producing anomalous quantum effects, *unless* those fields that satisfy appropriate differential equations. And such equations, written in the low-energy limit, miraculously coincide with the Einstein equations for the gravitational field, with the Dirac equations for the spinor fields, with the Maxwell and the Yang–Mills equations[22] for the gauge fields!

---

[20] The product of these three symmetry groups is at the ground of the so-called standard model of all fundamental interactions. The corresponding gauge fields are the photon (associated to the U_1 group), the carrier of electromagnetic interactions; the three vector bosons W^+, W^-, Z^0 (associated to the SU_2 group), the carriers of weak interactions; and the eight gluons (associated to the SU_3 group), the carriers of strong interactions.

[21] The sector of string states with spin higher than 2 is still a largely unexplored field of research (with the exception of the precious work of a few pioneers). The importance of such higher-spin configurations and the need for their further and deeper study have been repeatedly stressed in particular by Sagnotti (see for instance his recent review paper [47]).

[22] The Yang–Mills equations are the analogous of the Maxwell equations, written however for a non-Abelian gauge field. Unlike Maxwell's equations they are equations of non-linear type.

exactly as observed in Nature. There is no other theoretical scheme, to the best of my knowledge, able to make similar predictions.

Such a peculiar property of string theory probably represents the most revolutionary aspect of string models with respect to the other, more conventional, physical models, based on the notion of elementary particle: by quantizing the motion of a particle, in fact, we do not get any restriction on the evolution of the external fields in which the particle is embedded and with which it interacts. The equations of motion of those fields can be arbitrarily assigned and are chosen, as a rule, only on the basis of the available experimental information.

We can think, for instance, of the Maxwell equations for the electromagnetic fields, laboriously extracted in the nineteenth century from the empirical laws of Gauss, Lenz, Faraday, and Ampère. From a theoretical point of view it would be possible, in principle, to formulate sets of equations different from those of Maxwell but still preserving the electromagnetic gauge invariance (see, e.g., Sect. 2.3, footnote 21), the Lorentz invariance,[23] and all the other properties typical of the electromagnetic interactions. How to choose among all possible sets of equations?

If we are not aware of string theory (or we are not willing to use it), we can select the Maxwell equations and discard other possible choices only on the grounds of phenomenological motivations (agreement or disagreement with experimental observations). In context of string theory, instead, all possible alternatives to Maxwell's equations must be discarded a priori, as they would be inconsistent with the quantization of a charged string interacting with an external electromagnetic field. The same is true for the interactions of a string with an external non-Abelian gauge field, or gravitational field, or spinor field.

How do we get this remarkable conclusion?

We should recall, in this connection, that the string (and superstring) models are invariant with respect to the so-called conformal transformations (or local scale transformations[24]), whose action deforms the geometry of the string world-sheet without changing its dynamics. It is just because of this property that we can quantize a string (or a superstring) by choosing an adapted system of coordinate—the so-called conformal gauge—where the world-sheet geometry is flat, and we can also eliminate the longitudinal string oscillations, leaving as a complete set of independent degrees of freedom (satisfying the Virasoro constraints) only those modes describing oscillations transverse to the string itself. The conformal invariance thus plays a crucial role in the quantization process determining the spectrum of allowed physical states.

Let us then consider a "test" string, interacting with any one of the fields present in its quantum spectrum. For instance a string embedded in a curved space–time,

---

[23]This type of symmetry, which is at the core of the theory of special relativity, imposes on the electromagnetic equations to keep the same form in all inertial frames (whose coordinates are connected, indeed, by a Lorentz transformation).

[24]They are also called "Weyl transformations."

whose geometry is bent under the action of an external gravitational field, or a charged (open) string moving in the presence of an external gauge field.

The model describing interactions of this kind is called "sigma model" and is formulated so as to automatically respect conformal symmetry at the classical level.[25] However, for consistency, the same symmetry has to be respected also at the quantum level, given that the fields interacting with the string belong to the spectrum of quantum states obtained by exploiting precisely the condition of conformal invariance. This means, in other words, that we must impose on the quantized sigma model the absence of any "conformal anomaly."

The quantization process, on the other hand, often introduces expressions which are formally divergent. This occurs also in the case of the sigma model, which is to be regularized according to the standard rules of quantum field theory: what is required, in particular, is the addition of appropriate "counterterms" able to cancel the existing divergences. Such new terms, which are absent in the classical version of the model, in general are not conformally invariant, and thus introduce anomalies.

A computation of the conformal anomaly induced by these new terms shows, however, that the violation of the conformal symmetry explicitly depends on the external field interacting with the string, and that such a violation exactly reduces to zero when the external field satisfies a well-defined set of differential equations. It is precisely this set of conditions—that we must impose to prevent anomalies—that uniquely determines the dynamics of the considered field.

We should notice, at this point, that the contribution of the quantum counterterms is very difficult to be computed exactly. What is done, in practice, is to adopt an approximate approach, expanding the quantum corrections to the sigma model as an infinite series of terms,[26] whose relative importance is weighed by growing integer powers of the square of L_S (the typical size of a quantized string[27]). This means, in other words, that the various terms of the series are proportional to $L_S^2, L_S^4, L_S^6, \ldots$, and so on.

In the limit in which the string length L_s goes to zero we are lead back to the case of a point-like object, and the quantum corrections associated to the intrinsic extension of the string disappear. In that case there is no contribution to the conformal anomaly, and no constraint determining the equations of motion of the various fields.

If L_s is nonvanishing, on the contrary, the conformal anomalies produced by the series of quantum corrections must be canceled by imposing on the fields appropriate differential conditions. These conditions, like the quantum corrections, can be expressed as an infinite series of terms weighed by increasing (integer) powers of L_S^2. Higher powers of L_s will correspond, for obvious dimensional reasons,

---

[25]See for instance the textbooks [39–42] for an explicit and detailed introduction to such model.

[26]It is the usual perturbative expansion into a series of quantum *loop* corrections, referred, however, to the sigma model, i.e., to a two-dimensional field theory defined on the string world sheet.

[27]Concerning the correct physical interpretation of L_S see, in particular, Sect. 5.5.

to a higher number of derivatives appearing in the corresponding differential equations.[28]

In first approximation, the condition of conformal invariance imposes differential equations of the usual, second-order type (namely, equations containing the product of two differential operators). In second approximation the quantum corrections will introduce products of four differential operators. And so on, for approximations of higher and higher order, introducing always higher powers of the differential operator.

We can say, therefore, that the full quantization of the interaction of a string with an external field not only give us, in first approximation, the classical equations of motion for that field; it also provides us—as a series of increasingly accurate approximations in powers of L_S^2—with all quantum corrections to be applied in the regime where the field gradients are large (with respect to the distance scale L_S).

These higher-derivative corrections are typically associated to the string extension (indeed, they are absent in the context of ordinary quantum field theory) and are of crucial importance in the regime of large enough field gradients, i.e., strong enough forces. However, they are not the only corrections predicted by string models to the equations describing the evolution of the classical fields.

5.4.1   The Dilaton and the Topological Expansion

There are, in fact, other quantum corrections induced not by large values of the field gradients but by large values of the parameter g_S, controlling the intensity of all string interactions. In that case, also, there are quantum contributions to the equations of motion of the various fields, and also these new quantum contributions can be expressed as an infinite series of terms of higher and higher order, corresponding to increasingly accurate approximations.

The various terms of this series, as we shall see, can be associated to appropriate deformations of the string world-sheet introducing topologies of increasing complexity.[29] On the other hand, the various terms of the expansion are induced by the string interactions, hence their relative importance is weighed by growing powers of the square of the string coupling (g_S^2, g_S^4, g_S^6, ...).

It follows that there is a precise connection between the powers of the coupling g_S^2 and the topological level of the world-sheet surface (similar to the connection previously found between the powers of L_S^2 and the number of derivatives in the

---

[28]The operator representing partial derivative has dimensions of the inverse of a length. Given that the quantum corrections of order L_S^2 produce field equations containing the square of the differential operator, the quantum corrections of order L_S^4 will produce equations containing the fourth power of the differential operator, and so on.

[29]Topology is the science providing a quantitative description of the *global* geometric property of a space, such as compactness, connectedness, continuity, and boundary.

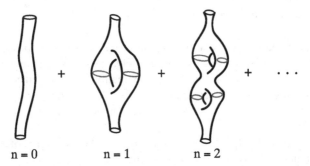

Fig. 5.5 Expansion of the world-sheet of a closed string into topologies of increasing level of complexity. The figure shows, in particular, a string freely propagating without any interaction ($n = 0$), the spontaneous production of a pair of virtual strings ($n = 1$), and the spontaneous production of two pairs of virtual strings ($n = 2$). The numerical value of n identifies the topological genus (or the number of holes) of the considered string world-sheet and also represents the order of approximation (or number of loops) of the corresponding quantum correction

field equations). And it is just because of this connection with topology that it will be possible to express the string coupling parameter as a function of the dilaton, and to interpret the dilaton as the field controlling the strength of all basic string interactions.

Let us consider, for instance, a closed string (or superstring). The propagation of this string in the external space–time describes a closed two-dimensional world-sheet surface which, in the absence of interactions, is of cylindrical type. Any intermediate portion of this world-sheet, interpolating between an initial and a final string configuration, has a compact geometric structure topologically equivalent to that of a sphere.

In the presence of quantum interactions, however, the initial string can spontaneously split into two strings, which eventually recombine to form a final state containing a single string (see Fig. 5.5). Such a process is analogous to the spontaneous production of a pair of virtual particles, a familiar process of quantum field theory occurring to first order of the so-called loop expansion of the quantum corrections.

If we take into account that process we can see, in the figure, that the intermediate portion of the world-sheet surface is no longer of cylindrical type, but it looks more like a ring-shaped "doughnut": its topology is no longer that of the sphere but that of the "torus," a geometric shape of topological genus $n = 1$. We should recall here that the topological genus is a property which characterizes the shape of an object by counting the number of "holes" (or "handles," as they are usually called in the technical language). Hence, a sphere has topological genus $n = 0$, a torus has topological genus $n = 1$, and so on.

In the regime in which the interactions are very strong, the quantum process of production and recombination of the string pairs can occur more and more frequently (see Fig. 5.5). We can then obtain world-sheets of increasing topological complexity, of genus $n = 2$, $n = 3$, etc. (i.e., with 2 holes, 3 holes, etc.).

Just like the genus 1 corresponds to a first-order correction (if referred to the series of quantum loop contributions), the higher-level topologies ($n = 2$, $n = 3$, and so on) will correspond to higher-order corrections (i.e., to quantum processes with 2 loops, 3 loops, and so on). In all cases, there is always a total and complete correspondence between the number of holes (i.e., the genus) of the topological expansion, and the number of loops of the quantum approximation.

We should now recall that the topological genus n of a given world-sheet can be obtained by computing the so-called Euler "characteristic integral"[30] χ. One finds, in particular, that $\chi = 1 - n$, and that the interaction of the string with the dilaton field—assuming, for simplicity, a constant dilaton—is precisely controlled by the product of the dilaton field ϕ times the Euler integral, i.e., by the term $\phi\chi$.

On the other hand, according to the laws of quantum mechanics, the probability amplitude of a string configuration described by a world-sheet surface of topological genus n (or, equivalently, of Euler characteristic χ) is inversely proportional to the exponential of the term describing the interaction of the given string with all the fields present in the string spectrum[31]: hence, it is inversely proportional also to the exponential of $\chi\phi$. Otherwise stated, it is directly proportional to the exponential of $-\chi\phi = (n - 1)\phi$, i.e., proportional to e^ϕ raised to the power n.

If we expand the total probability amplitude in a series of infinite contributions of increasing topological genus we thus obtain an expansion in integer powers of e^ϕ. But the different topological levels correspond, as already stressed, to the different approximation levels of the expansion in loops of the quantum corrections. And, according to the quantum mechanics, the various terms of the loop expansion are weighed by increasing powers of the coupling constant g_S^2.

We are thus lead to identify—at least in the case of a constant dilaton[32]—the string coupling constant with the exponential of the dilaton field, according to the simple relation $g_S^2 = e^\phi$. Such a relation, of crucial importance for the physical applications of string theory, is valid in this form only if the coupling is so weak that the quantum corrections can be expressed as a series of increasingly accurate approximations. Otherwise, the relation between the dilaton and g_S^2 still exists, but is more complicated.

In any case, it is important to stress that the effects of the world-sheet topology induce, in general, different couplings of the dilaton to the various fields, and different corrections to their equations of motion: this breaks the universality of the gravitational interactions and produces, in particular, scalar forces violating the principle of equivalence (as anticipated in Sect. 2.1.1). This effect is present at any

---

[30]It is defined by the integral over the whole world-sheet of a term proportional to the curvature of the world-sheet surface. The result is a constant number χ which is independent on the particular choice of coordinates, and is determined by the topological genus only.

[31]We are considering, in particular, the contribution of a given topological configuration to the so-called total "partition function" which controls the string propagation from an initial to a final state. Such a contribution is inversely proportional to the exponential function of the Euclidean action which includes all possible string interactions.

[32]If the dilaton is not a constant e^ϕ still plays the role of a *local* effective coupling.

level of approximation and tends to become more and more important as the value of the string coupling is growing.

5.5   A New Symmetry: "Duality"

There are two important symmetries which are present in string theory and which have no counterpart in the standard theories of fields and particles, as they have origin from the fact that strings are elementary but *spatially extended* objects. Both symmetries play a crucial role in the quantum version of the string models and are source of new and unexpected string properties.

One of these symmetries is the conformal invariance that we have discussed in the previous sections. This symmetry is closely related to the two-dimensional nature of the string world-sheet, and thus applies to elementary objects extended along *a single* spatial dimension. As we have seen, it is precisely this symmetry that, at the quantum level, determines the equations of motion of all fields present in the string spectrum.

The other symmetry, to be discussed in this section, is the so-called duality invariance.

This symmetry is present in string theory in various forms (duality of "type T," "type S," "type U"), and here we shall focus on the one that more explicitly refers to the string property of being an extended object: the duality of type T (also called "T-duality"), which becomes particularly evident when a string is quantized in a space–time containing compact dimensions.[33]

Let us immediately anticipate that, thanks to this symmetry, a string cannot distinguish—at the quantum level—a compact dimension of size R from another one of size L_S^2/R (where L_S is the typical length of a quantized string). The distinction between these two compact spaces is possible for a point-particle (of classical or quantum type), is possible for a classical string, but it is impossible to be made within the discrete energy levels of the spectrum of a quantized string.

How do we reach such an important conclusion? Its physical implications, as we shall see (also in the next chapter), are various and very interesting.

Let us take a simple example in which there is a single compact dimensions. Actually, the number of compact dimensions could be larger (as we have seen in the case of superstrings, which are vibrating in a ten-dimensional space–time); a single compact dimension, however, is already enough to introduce the basic ideas underlying duality symmetry. Let us assume, for simplicity, that this dimension has the shape of a circle of radius R, and consider a test body moving along such a compact, circular dimension.

---

[33]The letter T refers to the name "Target space," which is the name used in mathematics to denote the external space in which the string is embedded.

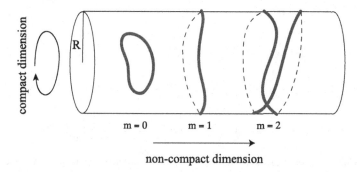

Fig. 5.6 A simple example of the two-dimensional space with one (horizontal) uncompactified dimension, and the vertical dimension compactified on a circle of radius R. A closed string can wrap many times around the compact dimension. In the figure we have shown (*from left to right*) three possible string states with "winding number" m corresponding, respectively to $m = 0$, $m = 1$ and $m = 2$. The winding number counts the number of times the string wraps around the circle

Let us take, first of all, a point-particle moving according to the laws of classical mechanics. The equations describing its motion depend, of course, on the radius of the circle: we have, for instance, that to complete a round trip it takes more time on a large circle than on a small one. We can thus say that the particle, from a classical point of view, "feels" the size of the circle and can physically distinguish compact dimensions of different radius.

The same conclusion is valid also in a quantum context: the energy spectrum of the particle, in fact, still depends on the radius of the circle, with the only difference (from the classical regime) that the energy levels are now discrete, instead of being continuous. The velocity of the particle along the compact dimension, in particular, takes values which are integer multiples of the inverse of the radius, i.e., values proportional to n/R, with $n = 0, 1, 2, 3, \ldots$. We thus recover the same spectrum of states typical of the so-called tower of Kaluza–Klein (see Sect. 2.3.1).

Consider then a closed string as our test body. If we consider its classical evolution in the presence of the compact dimension, the conclusion we reach is similar to that of the previous case: the classical string can physically distinguish spaces of different sizes. And this is done in two ways since the string, besides rotating along the circle (like the particle), can also wrap around the circle itself (see Fig. 5.6). Both states, i.e., the rotational states and the wrapping states (also called "winding" states) provide contributions to the total energy of the classical string which depend on the radius of the compact dimension.

The situation changes, however, when we move to the quantum regime.

The quantum spectrum of allowed energy levels obviously contains all possible dynamical contributions, and thus contains the energy associated to both the rotation and the wrapping of the string around the compact dimension. The kinetic contributions of the rotation are integer multiples of the inverse of the radius, hence (as previously mentioned) are of the type n/R, with $n = 0, 1, 2, \ldots$. The

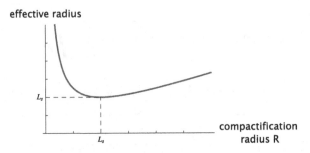

Fig. 5.7 Compact spaces of radius R, smaller than L_S, are equivalent, for what concerns the energy spectrum of a quantized string, to compact spaces of effective radius L_S^2/R, larger than L_S. The effective radius "felt" by the string—illustrated in this figure as the arithmetic mean of R and L_S^2/R—is always larger than (or equal to) the minimal value L_S

contributions of the winding energy, instead, are directly proportional to the radius and to the number m of windings: hence, they are of the type mR/L_S^2, with $m = 0, 1, 2, 3, \ldots$.

As the single contributions cannot be discriminated within a given energy level (what matters, for the spectrum, is the total energy only), it turns out that the quantum spectrum is invariant under the transformation[34]—called T-duality transformation—which exchanges between them the two integer numbers n and m and, simultaneously, inverts the compactification radius, interchanging $1/R$ and R/L_S^2.

This means that, in the regime of distances and energies where we must apply the laws of quantum mechanics, geometrical configurations of size R and of size L_S^2/R are physically indistinguishable for a string. And this implies, in practice, that the length scale L_S (typical of a quantized string) becomes the minimal scale of distances physically relevant in a string theory context (see Fig. 5.7).

It is important to stress that the appearance of a minimal length scale, in this context, is a direct consequence of quantum mechanics and not of the classical properties of a string. In a classical context, in fact, a string is characterized by a finite but arbitrarily small spatial extension: it is the quantization process which prescribes for the string a characteristic size whose square, L_S^2, is proportional to the Planck constant h, to the light velocity c, and is inversely proportional to the (classical) string tension.

In the classical limit in which the Planck constant goes to zero the minimal length L_S also vanishes, and the duality symmetry (which makes small and large distances indistinguishable) disappears. The length scale L_S thus represent, for a string, the analogous of the minimal radius of an atom (the so-called Bohr radius) obtained by quantizing an atomic system.

---

[34] As shown for the first time in the papers by Kikkawa and Yamasaki [48] and by Sakai and Senda [49].

It should be stressed that the property of duality invariance—illustrated here for the particular case of compact dimensions—can be extended so as to imply a physical equivalence between more generic distance scales, and not only compactification radii. This generalized invariance can be implemented even in the absence of compact dimensions, and even in the case in which the distances are time-dependent[35] (because the spatial geometry depends on time).

In these last cases, however, the duality symmetry is implemented in more complicated forms: the duality transformations, besides interchanging a distance and its inverse, also involve the dilaton (because the dilaton, present in the multiplet of fields described by the ground state of the spectrum, is unavoidably coupled to the graviton, and thus to the geometry of the external space in which the string is embedded).

The dilaton—or, better, the exponential function of the dilaton—represents, on the other hand, the string coupling g_S (see the discussion of Sect. 5.4.1). Hence, a transformation acting on the dilaton also transforms the coupling parameter: in particular, a change of sign of the dilaton leads to invert g_S. Considering this particular transformation we can then discover another type of symmetry, the so-called S-duality symmetry, corresponding to the invariance under the inversion of the string coupling (namely, under the transformation interchanging g_S^2 and $1/g_S^2$).

We have already noticed, in Sect. 5.3.3, that such a transformation relates type I superstrings and heterotic superstrings. More generally, combining various generalized forms of duality symmetries, we can relate all the five models of superstrings among them, in the sense that we can switch from one to another by applying suitable transformations acting on the fields, on the coupling constant and on the geometry.

This solves an old problem, arisen in the 1980 immediately after the birth of superstring theory: given that there are five possible models of superstrings (see Sect. 5.3), which is the "right" model, namely the most appropriate one to describe the world we live in?

The duality symmetry actually shows that this is a false problem. Being connected among them by duality transformations, the different models of superstrings do not correspond to different theories, but represent instead different physical regimes *of the same theory*.[36] It follows, in particular, that the five superstrings should correspond to different approximated version of a more fundamental model—the so-called "M-theory"—which has to be formulated in a space–time with $D = 11$ dimensions (11 is the highest allowed number of dimensions for a field

---

[35] As shown by the works of Tseytlin [50] and Veneziano [51].

[36] This result characterizes what is known as "the second superstring revolution," which took place in the mid 1990. There is also what is called "the first superstring revolution," which dates back to the 1980, and which corresponds to the transition from string theory intended as a theory of strong interactions (with a string length of the order of the nuclear radius, $L_S \sim 10^{-13}$ cm) to a supersymmetric string theory, intended as a unified theory of all fundamental interactions (including gravity, and with a string length of the order of the Planck radius, $L_S \sim 10 L_P \sim 10^{-32}$ cm).

theory of gravity which is supersymmetric and which does not include particles of spin higher than 2).

Our level of knowledge of the M-theory is currently very low.[37] We know, however, how to interpret the 11th dimension, given that superstrings oscillations need (for their quantum consistency) ten space–time dimensions only: in the M-theory context the 11th dimension, once compactified, can geometrically represent the intensity of the interactions among superstrings.[38]

The radius of compactification of the 11th dimension, in fact, turns out to be proportional to the exponential of the dilaton field, and thus proportional to the string coupling g_S^2. The strong interaction regime, where the full M-theory is to be applied without approximations, corresponds to large values of the radius, i.e., to geometric configurations in which the 11th dimension has a large, not negligible extension. The weak interaction regime is reproduced instead in the limit in which the compactification radius goes to zero, the effective space–time geometry tends to become ten-dimensional, and we can then apply the superstring models as a first-order (weak-coupling) approximation to the exact M-theory description.

It is worth noticing that the above geometric interpretation of the coupling g_S clearly displays the close connection existing between S-duality and T-duality. In the M-theory context, in fact, the inversion of the coupling constant (S-duality) corresponds to an inversion of the compactification radius of the 11th dimension, and thus represents a particular form of T-duality transformation.

The duality symmetry plays a crucial role in explaining the physical differences among the five types of superstrings, and in their interpretation as different perturbative approximations of a more fundamental theoretical scheme, formulated in 11 dimensions. Unfortunately, however, it seems unable to suggest solutions to what is currently regarded as one of the main conceptual problems of string theory: the so-called landscape problem.[39] What is meant by such a strange name?

It is meant that the five superstrings models, fully specified in their exact version valid at all energy scales, once required to make predictions in the low-energy approximation can produce about 10^{500} possible effective models of fundamental interactions, all different from each other. Really a huge number! And only one of these possible low-energy scenarios applies to the world in which we live.

Hence, the problem is that of moving from the exact version of the superstrings model to their low-energy limit, "landing" on the right effective model among all possible ones. String theory seems to give us no hint on how to choose the landing point and this, unfortunately, inevitably reduces its ability to make predictions. If we

---

[37]The M of the name has a number of possible interpretations: it may refer to Monster ("monster theory"), or to Mother ("mother of all theories"), or to Membrane ("membrane theory"). This last interpretation is due to the fact that, switching from 10 to 11 dimensions, and adding a new spatial dimension to a one-dimensional object like a string, one obtains a two-dimensional extended object: a membrane.

[38]As shown in a work by Witten [52]. It was just that work that, in practice, gave the start to the second string revolution.

[39]See, e.g., the paper by Bousso and Polchinski [53].

ask string theory how to describe a phenomenon that we can produce in our lab at low enough energy, the theory leaves us to choose among 10^{500} possible answers!

Such a situation is somewhat discouraging, and we can only hope, at present, that the future developments of string theory, or M-theory, will show us that the landscape problem is resolvable or that—as it often happens—the problem is ill-posed. After all, the available store of technical tools needed to mastering these new theoretical schemes is still largely incomplete and inadequate: in the presence of string theory we are a bit like a children pressing at random the keys of a computer, to see what happens.

However, even if (presently) useless with reference to the landscape problem, the duality symmetry—or, more precisely, the presence of the energy states associated to winding strings—suggests an answer to another important issue.

If we are living in a higher-dimensional space, and if the spatial dimensions are compact, as suggested by T-duality, then why *only three* dimensions are extended on extremely large scales of distance, while the other dimensions are characterized by a much smaller size (probably of the order of the string length L_S)? If our Universe has expanded since the Big Bang epoch, as predicted by the standard cosmological model, why this expansion has involved only three dimensions, without affecting the others?

In the following section we will see how the existence of states describing "winding" strings can provide an answer to these questions.

5.5.1 "Winding" Strings and Large Dimensions

The ability of strings of wrapping around the compact dimensions (see Fig. 5.6) represents not only a basic ingredients of the duality symmetry, but also one of the most typical and innovative aspects of string theory with respect to the (classical and quantum) theory of fields and particles.

The existence of the so-called winding states, with discrete, equally spaced energy levels, and with energies proportional to the radius of the compact dimensions, is a *unique* prediction of string theory. Other predictions, like supersymmetry, extra dimensions, and so on, are shared with other (more conventional) theoretical schemes.

If future experiments—as, for instance, the high-energy particle collisions of the LHC accelerator at CERN—should discover particles belonging to supersymmetric multiplets, or particles with the typical energy spectrum of the Kaluza–Klein towers, we could conclude that in Nature there is supersymmetry and there are extra dimensions, but we would not obtain any direct evidence in favor of strings (only an indirect support to their existence).

Should the experiments discover, instead, new particles with a spectrum typical of the winding states, we would obtain indisputable evidence that our physical world includes extended elementary objects like strings. Will we ever get an experimental result like that? We can only wait. In the meantime, however, we might wonder about new possible effects physically related to the presence of those winding configurations which are typical of strings and, more generally, of higher-dimensional extended objects.

A simple but interesting effect can be found in the context of the primordial cosmological models describing the very early epochs of our Universe.

Let us consider the Universe in an initial configuration of extremely high energy, characterized by a higher-dimensional geometric structure with all spatial dimensions compactified on really small scales of distance, of the order of the string length L_S. Let us suppose, also, that the Universe is filled with an extremely hot and dense gas of high-energy superstrings, propagating at relativistic velocities and repeatedly wrapping around the compact dimensions.[40]

Such a gas contributes to the comic energy density in two ways: through the kinetic energy due to the string motion, and through the winding energy associated to the wrapped strings. The first type of energy generates a positive effective pressure, similar to that of relativistic radiation, and its gravitational contribution tends to sustain the expansion of the spatial dimensions. The second type of energy, on the contrary, tends to favor the spatial contraction.[41]

Suppose now that the Universe starts to evolve in a phase of expansion: the winding energy—proportional to the radius of the compact dimensions—then necessarily grows and, as a consequence, sooner or later dominates. When dominating, the winding contribution can balance and rapidly overcome the expansion tendency, forcing the Universe to contract and to evolve back toward its initial configuration. As the radius decreases, however, the kinetic energy tends to become dominant again, stopping the contraction and restarting expansion. And so on.

Thanks to the opposite gravitational effects of the kinetic and winding contributions, a compact Universe, filled with a network of wrapped strings respecting the duality symmetry, is thus in a state of dynamical equilibrium where—apart from the periodic oscillations due to the alternation of the dominant forms of energy—the size of all dimensions is blocked to a value controlled by the initial compactification radius. It would seem impossible, in such a context, that three (and only three) dimensions can start an unbounded expansion, giving rise to the Universe that we are presently observing.

We should take into account, however, that at very high energies and temperatures the string gas tends to approach a symmetrical configuration containing the same number of strings wrapping around the compact dimensions in both orientations. The two different orientations correspond to states with opposite

---

[40]This possible initial configuration has been originally suggested and studied by Brandenberger and Vafa [54].

[41]For a direct check of the fact that the winding energy induces the contraction of the spatial geometry we should explicitly solve the cosmological equations describing the gravitational field produced by the considered gas of strings (see e.g., the paper by Tseytlin and Vafa [55]). Without resorting to such technical procedures, however, we can intuitively imagine that the strings wrapped around the compact dimensions behave as a "noose" which tends to tighten and to squeeze the space, counteracting its natural tendency to expansion.

"winding charges," and these states can annihilate each other when two strings come into contact, just as it happens for two particles of matter and antimatter when they collide.

It could be that, due to a process of mutual annihilation, all winding states present in the initial cosmological configuration gradually tend to disappear, leading to strings with kinetic energy only, and thus allowing the spatial geometry to expand. In that case, however, why should only three dimensions be able to escape the network of wrapped strings, expanding without limits? why should this not occur for *all* spatial dimensions?

The answer is simple. In order to annihilate each other, the wrapped strings must collide. If the number of spatial dimensions is too high, it is quite possible that such collisions will never occur, even if all dimensions are compact.

In order to explain this point let us take a simple example, considering two point-like particles moving around a single compact dimension (of circular shape, for instance). Unless their rotation velocities are exactly the same, and are oriented in the same direction, the two particles are doomed to collide, sooner or later. Let us suppose, however, that the particles are moving not on a circle but on a two-dimensional compact surface (for instance, a sphere): in that case they may never meet, even if their velocities are very different (it is enough, for instance, that they are moving along different parallel circles of the sphere).

Let us now consider two strings. Unlike particles, two strings wrapped on a sphere have a finite (and large) probability of colliding. The same is true if the two strings are moving not on a sphere but on a compact three-dimensional hypersurface.[42]

Iterating these arguments, and generalizing our examples, we may consider extended objects more complicated than strings (such as two-dimensional membranes or, more generally, p-branes, i.e., elementary objects extend along p spatial dimensions), wrapped around compact spaces of higher and higher dimensions. One then finds that the *maximum* number of compact dimensions in which the collision of two p-dimensional objects is unavoidable—with the consequent annihilation of their possible winding states—is given by $d = 2p + 1$.

For point-particles (without spatial extension, i.e., with $p = 0$) we then recover the result that the collisions are unavoidable only in a compact space with $d = 1$ dimensions (the circle of the previous example).

Strings, being one-dimensional objects, have $p = 1$, and their collisions cannot be avoided in compact spaces with 1, 2, or at most $d = 2 + 1 = 3$ dimensions. But the collision probability becomes negligible in spaces with a number of spatial dimensions larger than three! The winding states of the wrapped strings can thus meet, and completely annihilate each other, *only* in a three-dimensional subspace of a higher-dimensional Universe: here is the reason why *only three* spatial dimensions

---

[42]It is difficult, however, to get a graphic visualization of such a situation because, to do that, we should imagine the three-dimensional compact hypersurface embedded in an external four-dimensional space (which is impossible for the ability of our mind).

have managed to become free from the effects of the winding energy, and have been able to grow large and to expand on cosmic scales.

Within the remaining spatial dimensions—whether they are 6 as predicted by superstrings, or 7 as suggested by M-theory—wrapped strings have not the possibility of experiencing enough collisions: the winding states thus survive, and keep "bridling" the expansion tendency of the spatial geometry. Here is why these additional dimensions stay small, confined on a microscopic scale of distance of the order of L_S.

Actually, the situation could be more complicated than the one I have just described if, in addition to strings, the Universe contains other higher-dimensional extended objects, contributing to the total winding energy with their own wrapped states.[43] In fact, for a generic p-brane the winding energy is proportional to R^p, where R is the radius of the compact dimensions. When R starts growing, because the Universe is expanding, the objects with the largest values of p are thus the objects that become dominant first, and that control how many spatial dimensions are free to increase their size.

Let us suppose, for instance, that the space in which we live has a total number of 9 compact dimensions. All p-branes with p larger than 4 (or equal to 4) have no difficulty to intersect and to collide in a nine-dimensional space (for $p \geq 4$, in fact, we have $9 \leq 2p + 1$). The winding states of those p-branes can thus rapidly annihilate each other, hence they cannot prevent in any way the expansion of the nine spatial dimensions. The expansion can be stopped, instead, by the presence of p-branes with $p = 3$, $p = 2$, and $p = 1$.

With the growth of the compactification radius, and the annihilation of all winding states with p larger or equal to 4, the three-branes are the first ones to become the dominant source of cosmic energy, and to attempt blocking the cosmological expansion. They are not completely successful, however, because their winding states undergo a complete annihilation in a subspace with $d = 2 \times 3 + 1 = 7$ dimensions, leaving such a subspace free to expand (the two additional dimensions, on the contrary, keep confined).

The size of the seven-dimensional subspace can grow only until the two-branes come into play. When they become dominant the expansion is halted, again, except for a subspace with $d = 2 \times 2 + 1 = 5$ dimensions (which is the maximum number of dimensions where the two-branes can collide, and annihilate their winding energy).

But also the five-dimensional space cannot grow too large because, sooner or later, the Universe becomes dominated by the winding energy of the strings, with $p = 1$. The string winding states can efficiently self-annihilate, as we have seen, at most in $d = 3$ dimensions: hence, we are eventually left with a final geometric configuration where a three-dimensional subspace is expanding, while the other six dimensions are blocked. The only difference from the previous scenario (with strings but without branes) is that now the six extra dimensions have different values

---

[43]If this is the case, the primordial Universe is characterized by the presence of a gas of "branes." Such a brane-gas scenario is discussed, for instance, in a paper by Alexander et al. [56].

of their compactification radii, because their size becomes "frozen" (in pairs of two) at different epochs, corresponding to different phases of the gas of brane controlling the cosmological evolution.

We do not know, for the moment, whether the extra dimensions predicted by string theory really exist, whether they are compact or not, and whether the presence of wrapped string (or wrapped branes) may represent the correct explanation of their geometric configuration, so different from the one of the macroscopic three-dimensional space.

We have seen, however, that string theory—so difficult to be correctly applied in the context of low-energy physics—could prove to be particularly useful to explain, or predict, high-energy effects typical of primordial cosmology. In that context, string theory can also suggest new and—to some extent—revolutionary generalizations of the standard cosmological scenario, as we shall see in the next chapter.

Chapter 6
The Very Early Past of Our Universe

Have you ever wondered how the Universe was born, or, at least, how it looked billions and billions of years ago, in times very far from our present epoch? Was the Universe already existing in the very far past, and was it the same as today, or was it much different, or, maybe, had it not yet come into existence?

Modern cosmology tries to answer also questions like these, by applying the scientific method of physics: namely, starting from experimental observations, formulating theoretical models able to explain observations, and testing the predictions of those models with experiments of always increasing accuracy.

The model used to describe the present stage of our Universe is the so-called standard cosmological model, formulated (and repeatedly refined, in various steps) in the second half of the past century.[1] Like all physical models it is based on a number of assumptions, partly suggested by observations and partly by the mathematical methods adopted for its formulation.

We should mention, in particular, the assumption that the three-dimensional spatial sections of our Universe, on large enough scales of distance, have neither preferred positions nor preferred directions; the assumption that all forms of cosmic matter and radiation behave as perfect fluids, with negligible friction and viscosity; the assumption that the cosmic radiation is in a state of thermal equilibrium; the assumption that the presently dominating forms of matter and energy correspond to the "dark" components of the cosmic fluid (i.e., to components that are "visible" only through their possible gravitational effects). And, above all, the assumption that the gravitational interactions are well described by Einstein's theory of general relativity, at all scales of distances.

Using these (and other) assumptions, the standard cosmological model has collected a long and impressive series of successes. It has been able, for instance, to explain and to geometrically interpret the process of cosmic expansion, providing a quantitative description of the current dynamical state of the Universe. But not only that.

---

[1] See for instance the textbooks by Weinberg [2, 57] or by Gasperini [3] (for a textbook in Italian).

M. Gasperini, *Gravity, Strings and Particles*, DOI 10.1007/978-3-319-00599-7_6,
© Springer International Publishing Switzerland 2014

In fact, the model describes a Universe which expands and cools down evolving from an initial state characterized by an infinite concentration of energy: the so-called "Big Bang" singularity. If we move back in time, starting from the present epoch, we find in the past states of increasingly high temperature and density, in a Universe dominated by a very hot gas of relativistic radiation. The standard model thus predicts the existence, in the past, of the environmental conditions appropriate to give rise to the nuclear reactions producing the basic chemical elements (hydrogen, helium, and so on), through the so-called nucleosynthesis process. In addition, such a thermal cosmological history explains the origin of the relic background of microwave radiation that we can still observe on cosmic scales.[2]

In spite of these important achievements, the standard cosmological model has been seriously put in trouble two times.

The first time was at the beginning of the Eighties, when the astrophysics community started to investigate the problem of the origin of the agglomerates of cosmic matter and of the small (but finite) inhomogeneities and anisotropies present in the relic cosmic radiation.[3] How did such fluctuations arise? We are referring, in particular, to the temperature fluctuations of the cosmic radiation and—most important—to the density fluctuations responsible for the concentration of the cosmic matter and for the subsequent growth of the cosmic structures (stars, galaxies, clusters) that today we are observing. Indeed, no temperature fluctuation and density fluctuation should exist, on cosmic scales, if the Universe were exactly homogeneous (i.e., the same at all points) and isotropic (i.e., the same along all directions), as prescribed by the standard cosmological model.

But, besides the above problem, there are further problems associated with the very "special" type of geometry assumed by the standard model.

If we go back in time, for instance, we find that the Universe is concentrated in regions of space of smaller and smaller size and that the cosmic gravitational field becomes stronger and stronger, so that the curvature of the space–time geometry—according to the Einstein equations—is always larger, and grows without limit.

---

[2]It is a background of electromagnetic radiation, filling the whole space with an almost uniform distribution of very weak intensity. Detected for the first time by Penzias e Wilson [58], is characterized by a spectral distribution of Planckian type, typical of radiation in thermal equilibrium. Its temperature gradually decreases as the Universe expands, and its present value is about 2.7 K. According to the standard model, the cosmic background of relativistic radiation should also contain a component of relic neutrinos in thermal equilibrium, at a temperature slightly larger than that of photons (see e.g., [2, 3, 57]). However, because of its very faint intensity, such a neutrino component has never been (so far) detected.

[3]Direct observations tell us that the temperature of the cosmic radiation may vary from point to point, deviating from its average value (2.7 K) with small percent variations not larger than about one part in a hundred thousand. Such tiny inhomogeneities have been directly measured for the first time by the satellite COBE [59] in the Nineties, and later (with always increasing accuracy) by the WMAP experiment in the first decade of this century. Currently, we are obtaining even more precise information thanks to the data collected by the PLANCK satellite, whose first results have been released on March 2013.

And yet, the three-dimensional space stays nearly flat, described by a geometry of Euclidean type! How can it happen? This is the so-called flatness problem.

Another problem is the so-called horizon problem. Consider the portion of space that we are presently able to observe: a very huge region of space, extending around us with a radius slightly below 14 billion light years. According to the standard cosmological model such a region, in the past, was much smaller than today. And yet, it was so large that even a light ray emitted at the very beginning (in principle, at the Big Bang time) would have not had enough time to cross it completely!

This means that—according to the standard model—the different portions of space that today we are observing have not had enough time, in the past, to exchange signals, hence have not had any possibility to physically interact among themselves (if—as we believe—no signal and no interaction can propagate faster than the speed of light).

Nevertheless, all portions of space presently observable are extremely similar to each other: the same average density, the same average temperature, and so on. It is hard to believe that this occurs by chance. So, which mechanism has enabled such regions to communicate among themselves, overcoming the causal "horizon" imposed by the existence of a limiting velocity?

These (and other) problems, arisen during the "crisis" of the Eighties, have been solved by modifying the original standard model, and introducing at some very early epoch a new type of cosmological phase, called "inflation." During this new phase the three-dimensional space "inflates" at an exponentially accelerated rate, undergoing a gigantic growth of its volume in a very small fraction of time.[4] Such a process is able, on one hand, to trigger the growth of the macroscopic inhomogeneities characterizing our present Universe[5]; on the other hand, it also automatically solves[6] the horizon problem, flatness problem, and so on.

The second, important crisis of the standard cosmological model is the one of the late Nineties (that we have already introduced in Sect. 3.2). The discovery that our Universe has (only recently) entered a phase of accelerated expansion[7] has produced a further modification of the original model. The required change amounts to the

---

[4]This idea, originally proposed by Guth [60], was later developed and refined by many other authors. For a deeper discussion of the various aspects of the inflationary regime see, e.g., the textbooks by Kolb and Turner [61] or by Gasperini [3] (for a textbook in Italian).

[5]The inflationary phase does not produce new intrinsic inhomogeneities, but simply amplifies the tiny quantum fluctuations unavoidably present in the matter fields and in the cosmic geometry.

[6]The inflationary phase, in fact, "pumps up" the sizes of those spatial domains which were small enough to have allowed reciprocal interactions, and makes them larger than the horizon size typical of that epoch (i.e., larger than the maximum distance a light ray would have had time to travel up to that moment). See Fig. 6.1.

[7]Such an accelerated phase is qualitatively similar, from a dynamical point of view, to the inflationary phase that the Universe has (or should have) experienced in its very early epochs. The typical acceleration of the inflationary regime, however, is by far more intense than the current cosmic acceleration.

introduction of a new type of "dark" energy, which—in its simplest form—can be conveniently represented as a cosmological constant Λ.

With the inclusion of an appropriate cosmological constant we obtain the so-called ΛCDM model,[8] or "concordance model," which represents the current version of the standard model able to reconcile (within the limits of the experimental errors) all available cosmological data into a single and coherent theoretical scheme. Such a model aims at providing a complete description of the history of our Universe, starting from the present accelerated epoch and going back in time as much as possible, through the phase dominated by the dark matter, the phase dominated by the radiation, down to the epoch of primordial inflation.

The inflationary regime predicted in the context of that model, however, cannot be extended towards the past for an infinite (or arbitrarily long) time interval.[9] If we go back in time down to sufficiently early epochs, we find that the inflationary phase of the concordance model must have a beginning at some given instant of time. Before that time we find the Universe in an extremely hot, dense, and curved primordial state—an ultimate concentrate of matter and radiation at extremely high energy and temperature.

This means, in other words, that before starting inflating the Universe was quite close to the Big bang epoch (see Fig. 6.1). Namely, quite close to the instant of the huge cosmic "deflagration" which—according to the standard model, even taking into account the inflationary corrections—gave rise to all species of matter and energy that we observe today, and also marked the origin of the space–time itself. According to the standard model, in fact, the Big Bang corresponds to a mathematical singularity where the energy density and the space–time curvature blow up to infinity: beyond that point, any physical model or theory ceases to be valid.

If we turn back to the questions posed at the beginning of this chapter (how did the Universe begin? what did it look at the beginning? was it always existing?) we can thus say that the answer provided by the standard cosmological model is quite precise and rather "drastic." The Universe—tell us the standard model—was born from an initial singularity about 14 billion years ago,[10] and before that epoch there

---

[8]The acronym CDM means Cold Dark Matter, and the name ΛCDM emphasizes the fact that, in this model, the cosmic gravitational field is currently generated by two main sources: cold (i.e., non-relativistic) dark matter and a cosmological constant Λ.

[9]As discussed, for instance, in the paper by Borde, Guth, and Vilenkin [62].

[10]We should keep in mind that such a number (14 billion years), like any other number expressing the duration of a time interval, is always to be referred to some particular observer and some particular choice of the parameter used to measure the time coordinate. In this case the 14 billion years are referred to the so-called cosmic time parameter, which is the time coordinate used by a "comoving" observer, i.e., by an observer placed at a fixed spatial position which gets "carried away" unresisting through space–time by the expansion of the cosmic geometry. What is important, anyway, in our context, is that the temporal distance between the Big Bang and the present epoch is *finite*: hence, according to the standard model, our Universe is not an "infinitely old" creature (namely, it has not existed forever).

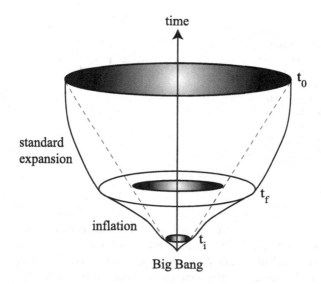

Fig. 6.1 A space–time diagram illustrating the expansion of a portion of cosmic space from the Big Bang to the present epoch t_0. The *vertical axis* represents the time coordinate and the *horizontal planes* correspond to spatial sections of the Universe at constant values of the time coordinate. The inflationary phase (characterized by exponential expansion) extends from an initial time t_i to a final time t_f. The subsequent phase describes the evolution predicted by the standard cosmological model, from the time t_f until now. The *dashed lines* illustrate how the size of regions connected by light signals (represented in the figure by the shaded areas) changes in time. The *solid curves* show, instead, the overall growth of the spatial volume due to the expansion of the cosmological geometry

was nothing of what is part of the Nature that we are presently observing (or, at least, nothing that may be accessible to our scientific investigation).

The discussion on the primordial Universe, at this point, could stop here, unless we recall that we cannot take too seriously the predictions of the standard cosmological model when we are too close to the Big Bang epoch. Why? For a very good reason: because it is a model based on the theory of general relativity, which is a *classical* gravitational theory, also valid in the relativistic regime, but certainly *not valid* in the *quantum* regime.

All classical theories, in fact, are characterized by a limited validity domain: they are valid until the so-called action S of the theory turns out to be sufficiently large with respect to the elementary "quantum of action" (or Planck's quantum) h. The action, in this case, is a mathematical quantity which controls the effective intensity of the physical processes that we are describing, taking into account both the distances involved and the duration of those processes.

Fig. 6.2 Evolution in time of the curvature of the Universe according to the standard cosmological model. As the Universe approaches the initial singularity (Big Bang) the Hubble radius L_H tends to zero, and the space–time curvature (which is proportional to $1/L_H^2$) goes to infinity

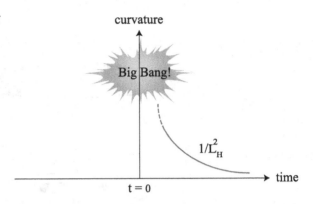

In our case, if we compute the action of general relativity[11] for the cosmic gravitational field predicted by the standard model, we obtain a number which is directly proportional to the square of the space–time curvature radius (or Hubble radius) L_H, and inversely proportional to the square of the Planck length L_P. The Planck length is a constant, but the Hubble length L_H depends on time and, in particular, grows as the Universe expands. Hence, also the action grows with the cosmic expansion.

Today the Universe has a very wide spatial extension, the Hubble length L_H is very large, and the gravitational action S thus easily satisfies the condition $S \gg h$ which characterizes the classical regime, and which ensures the validity of general relativity and of the standard cosmological model.

If we go back in time, however, we find a Universe which becomes more and more curved and concentrated (see Fig. 6.2): the Hubble radius L_H becomes, in proportion, smaller and smaller, and eventually tends to zero when the Universe approaches the initial time of the Big Bang. In that regime it is clear that also the action S (which is proportional to L_H^2) tends to become arbitrarily small: the condition which guarantees the validity of a classical model like the standard model is thus necessarily violated, at some time, before reaching the Big Bang. From that time on, the Universe enters a regime that requires the use of quantum physics.

Hence, in order to provide a reliable description of the very early cosmological epochs, we should not apply general relativity but a theory of gravity valid in the quantum regime. Does such a theory exist?

String theory, as we have seen in Chap. 5, not only *can* describe the gravitational interaction, but also *must* include such interaction, in the quantum context, in order to be a complete and formally consistent theory. In addition, string theory is valid for all interactions at arbitrarily high energy scales, hence it can be correctly applied to describe the Universe even in epochs arbitrarily close to the Big Bang singularity.

---

[11]The action for the gravitational field, in the theory of general relativity, is determined by a geometric quantity called "scalar curvature," integrated over the whole space–time region we are considering, and divided by the Planck length squared.

In those epochs, characterized by extreme physical conditions, the equations we obtain from string theory to describe the gravitational forces are different from the corresponding equations of general relativity and of the standard cosmological model. It makes thus sense to ask the question: what's new from string theory about cosmology? In particular, what's new about the very early epochs and the beginning of our Universe?

6.1 String Cosmology

Among the many innovative aspects of string theory, two of them, in particular, could play a relevant role in the formulation of a consistent cosmological scenario.

The first aspect concerns the duality symmetry that we have illustrated in Sect. 5.5. If such a symmetry is to be respected (even at the approximate level) by the cosmic gravitational field, then any cosmological phase with *decreasing* curvature, describing a Universe which expands from $t = 0$ for all positive values of the time coordinates (see Fig. 6.2), must be associated to a "twin" cosmological phase with *growing* curvature, describing a Universe which expands for all negative values of t, up to $t = 0$.

If we apply this condition to the present cosmological phase, described by the standard model and occurring after the Big Bang epoch, we find that such a phase should be preceded in time by a "dual" phase, with properties almost specularly symmetric to those of the standard cosmological phase, occurring *before* the Big Bang epoch (see Fig. 6.3). In view of their temporal locations, it is natural to call these two phases with the explicit names of *pre-big bang* and *post-big bang*.[12]

Both the phases illustrated in Fig. 6.3 are characterized by a Hubble radius which goes to zero (and a curvature which goes to infinity) as the Universe approaches the Big Bang, at the time $t = 0$. This time—as clearly shown in the figure—corresponds to a *future* singularity for the pre-big bang phase, and to a *past* singularity for the post-big bang phase that we are currently experiencing.

If this were the actual situation, then the two phases would be disconnected by a space–time singularity, with no chance of merging together into a single, coherent model of cosmological evolution. Even if the pre-big bang phase exists (or, better, has existed) in the past history of the Universe, it would remain confined forever in a space–time region inaccessible to our physical experience: no (direct or indirect) physical signal, no radiation, no interaction (no matter how strong) of pre-big bang origin could ever be able to cross the barrier of infinitely high curvature, infinitely high energy, and reach us in the post-big bang epoch.

---

[12]The cosmological "pre-big bang" scenario, suggested by string theory, has been formulated and discussed in the works by Gasperini and Veneziano (see for instance [63]). See also the book [64] for a detailed but qualitative introduction to the pre-big bang scenario, and the textbook [42] for a more technical discussion.

Fig. 6.3 In a model respecting the dual symmetry of string theory, the phase of standard cosmological evolution and decreasing curvature, following the Big Bang, should be always associated to a phase of growing curvature occurring before the Big Bang

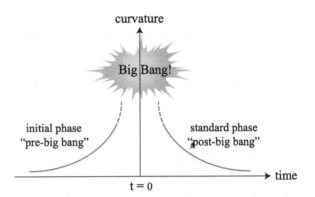

It is at this point, however, that another crucial aspect of string theory comes to our rescue: the existence of a minimal length L_S which, as we have stressed in Sect. 5.5, controls the typical size of a quantized string.

Distances and sizes smaller than L_S, for a string, have no physical meaning. It follows that a cosmological scenario consistent with string theory should be characterized by a Hubble radius L_H *always larger* than (or equal to) this minimal distance scale.

But if L_H can neither go to zero nor become arbitrarily small, then the space–time curvature (which is proportional to $1/L_H^2$) cannot become arbitrarily large or infinite! When the Hubble radius L_H reaches the limiting value L_S, the curvature reaches the maximum allowed value $1/L_S^2$ and, from that moment on, can only evolve in two ways: it can either stabilize at that value, or start decreasing toward lower curvature states. Hence, in the context of string cosmology,[13] the Big Bang singularity predicted by the standard model, and sharply localized at a given epoch (say, $t = 0$), tends to disappear, and to be replaced by a phase (possibly extended in time) of extremely high, but *finite*, maximal curvature: the so-called "string phase."

By combining the dual symmetry with the existence of a minimal length scale, string theory thus suggests a completion of the standard scenario which extends the physical description of the Universe back in time beyond the Big Bang and, in principle, back to infinitely remote values of the time coordinate (see Fig. 6.4). The Big Bang era is still there but, in this new context, it is deprived of the almost "mystical" role of initial singularity: it becomes, much more simply, the epoch of transition between the initial, growing curvature regime and the final, decreasing curvature, standard regime.

In such a context the initial state of the Universe is no longer localized at $t = 0$ but has been moved to the unimaginably distant past, and is located at an infinite temporal distance from our epoch. It corresponds to a so-called "asymptotic state" of string theory, namely to a limit state that is never fully achieved: it is only realized

---

[13] With the name "string cosmology" we will denote, generically, a cosmological scenario based on (or at least inspired by) string theory.

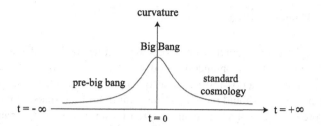

Fig. 6.4 Time evolution of the curvature of the Universe for a typical model of string cosmology. The Big Bang epoch does not correspond anymore to a singularity, i.e., to a state on infinite curvature, but to a phase of maximal, finite curvature. The model can be extended in time without limits, in principle to infinity, towards both the past and the future direction

in an approximated way, with an approximation which is always better as we go farther and farther back in time, towards the limit $t = -\infty$.

The asymptotic state obtained in this limit is the so-called string perturbative vacuum. As also suggested by Fig. 6.4, such a state turns out to be, in many respects, a sort of specularly-symmetric version of the final state that could be reached in the asymptotic future if the Universe would keep expanding forever. It is, in fact, an initial state with negligible curvature and energy density, extremely flat, cold, and empty: a configuration really different from the hot, extremely curved, concentrated, explosive initial state predicted by the standard cosmological model!

The only (but important) asymmetry, possibly existing between the initial and the final state of the above string cosmology scenario, and not illustrated in Fig. 6.4, is related to the strength of the fundamental interactions.

The effective intensity of all fundamental interactions, in fact, is controlled by the string coupling parameter g_S^2 (as discussed in Sect. 5.4.1). Such a parameter tends to zero in the asymptotic initial state, since the string perturbative vacuum is characterized by arbitrarily small interactions. However, during the phase of pre-big bang evolution, the value of g_S^2 is doomed to grow together with the space–time curvature, leading the Universe towards a phase of strong interactions, expected to be reached before the beginning of the standard, post-big bang regime.

The string coupling would keep growing even in the subsequent epochs, were it not properly stabilized in a range of values compatible with present observations. In any case, the growth of the coupling spontaneously breaks the symmetry otherwise existing between the past and the future asymptotic states of the cosmological model illustrated in Fig. 6.4.

In addition, the growth of the string coupling implies that the pre-big bang to post-big bang transition takes place in a phase characterized not only by large curvatures but also by strong interactions. In that context, as we shall discuss in the next section, the Universe can easily become filled with strings or, more generally, with extended objects like p-branes (if the space is higher-dimensional). The pre-big bang scenario, suggested by string theory, thus naturally leads us to consider the possibility of—even more exotic—"brane cosmology" scenarios.

6.2   Brane Cosmology

During the high-curvature and strong-coupling regime, typical of the end of the pre-big bang phase, the spontaneous production of branes from the vacuum is expected to be not only a possibility, but also a highly probable process. Why?

To explain why, we should recall that the evolution of the space–time geometry during the pre-big bang phase is characterized by an accelerated expansion rate: it is thus an evolution of inflationary type.[14] Inflation, on the other hand, can efficiently amplify all kinds of quantum fluctuations, whether they are associated with the geometry or with the various fields present in the cosmological model. The amplification of such fluctuations can be described, from the point of view of quantum field theory, as an effective productions of pairs of particles[15] from the vacuum.

In addition to elementary particles, however, string theory also predicts the existence of elementary extended objects: the p-branes that we have already introduced and discussed in Sect. 5.5.1. The value of p—i.e., the number of intrinsic dimensions of the extended object—obviously cannot be larger than the number of dimensions available in the external space: in particular 10 if we are using M-theory, or 9 if we are considering superstring theory, or even a smaller number, if we are assuming that a few dimensions have been compactified, and are decoupled from the inflationary dynamics.

In any case, the mechanism of the inflationary amplification leading to the production of particles (i.e., 0-branes) from the vacuum can also produce, in the same way, p-branes, with a number p of intrinsic dimensions different from zero. These extended objects must be produced in pairs like the particles, and the energy required for their production depends on their "tension," namely on the mass per unit of the p-dimensional spatial volume of the p-brane. The tension, in general, turns out to be inversely proportional to the string coupling g_S.

At the beginning of the pre-big bang phase, when the Universe is still quite close to the regime of the string perturbative vacuum, and is characterized by a very small value of g_S, the tension of the p-branes turns out to be very high: hence their production requires a big fluctuation of the vacuum energy density. The spontaneous production is possible, in principle, but very unlikely, in practice.

Toward the end of the pre-big bang phase—i.e., close to the phase of maximal curvature, where the string coupling g_S becomes strong enough—the tension of the p-branes turns out to be much smaller, instead, and a very little energy is

---

[14]Like the inflationary phase described at the beginning of this chapter, with the only difference that the space–time curvature is growing instead of being decreasing, since inflation occurs *before* the Big Bang.

[15]To be consistent with the conservation of the total charge, of the total angular momentum, as well as with all the existing physical conservation laws, particles have always to be produced in pairs: each produced particle must be associated to a produced antiparticle with opposite charge, opposite angular momentum, and so on.

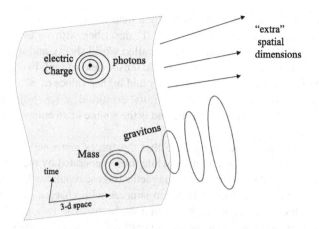

Fig. 6.5 Two possible sources of interactions localized on a three-dimensional brane (for simplicity, the figure illustrates only one of the three spatial dimensions of the brane). On the top of the figure we have a charge, source of electromagnetic field: the associated electromagnetic waves (or photons) are strictly confined, and can propagate only on the brane space–time. On the bottom we have a mass, source of gravitational field: the associated gravitational waves (or gravitons) can leave the brane space–time and propagate through all the available spatial dimensions, inside or outside the brane

needed to produce a pair of these objects. Here is why, in that regime, branes are easily produced and the higher-dimensional space–time tends to be filled with extended objects, interacting among themselves and wrapping around the compact dimensions (see Sect. 5.5.1 for the cosmological effects of the wrapping branes).

Let us now consider the possible interactions among branes. As we have seen in Sect. 2.4 for the case of Dirichlet branes, the fundamental interactions tend to be confined on the branes themselves: in fact the charges, sources of these interactions, are localized on the ends of the open strings, and these ends, in their turn, are rigidly constrained to move on a brane.

We know, however, that there is an exception. The gravitational force is carried by closed strings, and closed strings can propagate also outside the brane containing the gravitational sources, spanning all available spatial dimensions (see Fig. 6.5). The branes and antibranes produced in pairs by the regime of pre-big bang inflation can thus gravitationally interact among themselves through the mutual exchange of closed strings.

The effective gravitational force arising in this way, according to string theory, has various components. Besides the usual symmetric tensor field, representing the graviton, and a scalar field representing the dilaton, it also contains an antisymmetric tensor field representing the so-called forms of Kalb-Ramond.

This last contribution generalizes to the case of p-dimensional extended objects the usual vector field of forces associated to the point-like charges. Let us recall, indeed, that a point-like object (with $p = 0$) describes with its evolution a one-dimensional trajectory in space–time (the so-called world line), and, if charged, it is a source of interactions represented by a field of vector type (i.e., by a tensor field of rank one[16]). A string (with $p = 1$) describes with its evolution a two-dimensional surface in space–time (the so-called world sheet), and it is a source of interactions represented by an antisymmetric tensor field of rank two (the so-called Kalb-Ramond axion). And so on, for higher and higher values of p.

In general, a p-brane describes with its evolution a $(p + 1)$-dimensional hypersurface in the external space–time and is the source of an antisymmetric tensor field of rank $p + 1$.

What is important, in our context, is that the forces generated by the graviton and dilaton fields are always attractive, while those generated by the antisymmetric tensor fields are similar to the electromagnetic forces: repulsive between sources with the same charges, attractive between sources with charges of opposite sign. In particular, they are repulsive between two branes (or two antibranes), and attractive between a brane and an antibrane (carrying opposite charges with respect to the Kalb-Ramond interactions).

If we take now a system of two identical, static, and parallel branes, initially arranged in a perfectly symmetric state called BPS configuration,[17] we find that the forces of attractive type are exactly balanced by those of repulsive type, and the net result is a vanishing interactions. If we take, instead, a brane and an antibrane, then the mutual gravitational interaction is attractive and always non-vanishing, quite irrespectively of their initial configuration.

Because of their reciprocal attractive interaction, branes and antibranes (copiously produced by the phase of pre-big bang inflation) will unavoidably tend to collide (see Fig. 6.6). Their collisions become more and more frequent, and more difficult to avoid, as the Universe comes closer and closer to the phase of maximal curvature which anticipates the transition to the post-big bang era.

If—as suggested by the so-called brane-world scenario, see Sect. 2.4—our three-dimensional macroscopic Universe is nothing but one of these branes (in particular, a 3-brane) embedded in an external, higher-dimensional space, then it could be that it was just the collision of our "brane-Universe" with an anti-brane to simulate the Big Bang explosion, and to trigger the transition from the pre-big bang phase to the phase of standard cosmological evolution.

---

[16]The rank of a tensor is given by the number of indices which characterize its explicit representation, and "counts" the number of its components. For instance, a tensor of rank 1 is represented by an object with a single index: A_μ. A tensor of rank 2 is represented by an object with two indices: $F_{\mu\nu}$. And so on.

[17]From the initial of the names of Bogolmon'y, Prasad, and Sommerfeld.

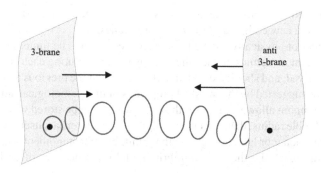

Fig. 6.6 Branes and antibranes tend to collide because their mutual gravitational interaction is always attractive (like the electric force between a charged particle and the corresponding antiparticle). In particular, the primordial Big Bang explosion might have been simulated by the collision of an anti-3-brane with the 3-brane corresponding to our macroscopic Universe

6.2.1   The "Ekpyrotic" Scenario

The model of Big Bang as a collision of branes was originally suggested in the context of the so-called ekpyrotic scenario,[18] inspired by string theory and based, in particular, on its M-theory generalization. In that scenario, however, there are no collisions of branes and antibranes, but only the collision of the two 3-branes marking the opposite boundaries of our macroscopic Universe, represented as a five-dimensional space–time (the additional six spatial dimensions, required by M-theory, are assumed to be compactified and characterized by a geometry of Calabi-Yau type).

In the context of the ekpyrotic scenario, the fifth dimension of the space–time—i.e., the spatial dimension orthogonal to the three-dimensional boundary hypersurfaces—is continuously contracting as the two boundaries approach each other: the size of this dimension goes to zero at the collision epoch, and then expand (with a sort of "bounce") after the collision. The distance between the two boundaries, however, never goes to infinity because of their mutual interaction: sooner of later the fifth dimension stops expanding, stays in balance for a moment, and then shrinks back again.

We are thus led to consider, in this context, a "cyclic" cosmological scenario[19] in which the two branes at the edges of the space–time periodically approach each other, collide, move away, then get close again, still collide, and so on. The whole process would seem to be able to repeat itself at will.

If this picture is taken seriously, it follows that today we are living in one of the (many) possible phases of standard cosmological evolution, which occurs after one

---

[18]See for instance the paper by Khoury, Ovrut, Steinhardt, and Turok [65]. The name of this scenario comes from the ancient Greek language, and denotes something which is "emerging from fire" (with reference to the old myth of the Arabian Phoenix).

[19]As suggested in the paper by Steinhardt and Turok [66].

of the (many) possible Big Bang epochs, waiting for a new contraction of the fifth dimension and a new collision of the boundary branes.

The number of physically allowed cycles, however, could be finite. According to the laws of thermodynamics, in fact, the entropy produced during each cosmological cycle is conserved, and should be added to the entropy of the previous and following cycles. If—as suggested by the physical properties of the cosmological horizons[20]—there is a maximum allowed value for the entropy that can be stored within a spatial volume of Hubble radius, then a scenario which cyclically repeats itself should bring the Universe, sooner or later, to saturate that limit. From that moment on the cosmic system should reach the thermal equilibrium, and the sequences of cycles should stop there.

Even if this is the case, the number of allowed cycles is still extremely large: given that the maximum entropy compatible with the present cosmological horizon[21] is a number of order 10^{122}, and that the entropy associated to the phase of standard evolution[22] is a number of order 10^{90}, it follows that one needs about 10^{32} cycles to reach the limiting state of thermal equilibrium!

In addition, a physical observer confined inside a single cycle (like us) should be unable to distinguish which cycle is experiencing: in fact, he is only aware of the entropy produced by that cycle, hence his epoch appears to him exactly the same as the corresponding epoch of the preceding and following cycle.

Inside each cycle, and during the phase preceding the collision of the branes, the space–time curvature grows like in the pre-big bang phase introduced in Sect. 6.1. The ekpyrotic scenario, however, is radically different from the pre-big bang scenario because, before the collision of the two branes, the strength of all interactions (controlled by the coupling constant g_S) *decreases* in time, instead of growing like in the pre-big bang phase. In addition, the pre-big bang phase is characterized by an accelerated expansion of inflationary type, while the ekpyrotic pre-collision phase is characterized by accelerated *contraction*.

The typical problems associated with the standard cosmological evolution (i.e., the flatness problem, horizon problem, etc.) are solved in the ekpyrotic scenario thanks to the presence of the contracting phase, with no need of introducing an inflationary regime of conventional type.[23] Inflation, however, is not incompatible with brane cosmology, and can be implemented even in its most conventional version, as shown by the model of brane cosmology that will be introduced in the next section.

---

[20] See for instance the paper by Goheer, Kleban, and Susskind [67].

[21] Such a maximum entropy is determined by the area of a spherical surface of Hubble radius L_H, measured in units of Planck length L_P. Hence it is a number of order L_H^2/L_P^2.

[22] The entropy density of the thermal radiation is proportional to the cubic power of its temperature T_H. The entropy of the radiation stored inside a cosmological volume of radius L_H is thus a number of order $T_H^3 L_H^3$.

[23] Such a difference, as we shall see below, has non-trivial impact on various properties of the produced gravitational radiation.

6.2.2 Brane–Antibrane Inflation

In order to reproduce a phase of standard inflationary evolution through the mechanism of the interacting branes we must reconsider the model illustrated in Fig. 6.6, i.e., the model where the 3-brane representing our Universe is experiencing the attractive gravitational force of a cosmological "twin" antibrane.

Let us suppose that the two branes are embedded in a higher-dimensional space with a number n of extra spatial dimensions, suitably stabilized in a compact configuration. There is no cancellation among the various components of the gravitational force, and the spatial distance Y between brane and antibrane play the role of a scalar field controlling the net intensity of their mutual attraction. The potential energy associated to this interaction, taking into account all contributions (of the graviton, of the dilaton, of the Kalb-Ramond forms), turns out to be proportional to Y^{2-n}.

Let us consider, in particular, a model based on superstring theory and formulated in $D = 10$ space–time dimensions. The total number of spatial dimensions is $D - 1 = 9$, and the number of "extra" dimensions, external to the 3-branes, is thus $n = 9 - 3 = 6$. The potential energy then varies with the distance between the branes as $1/Y^4$.

The presence of the antibrane introduces a new effective interaction on the 3-brane corresponding to our Universe. Such an interaction is represented by the scalar field Y and is characterized by a potential energy which increases as the two branes approach each other. Most importantly, this potential energy is able to sustain the accelerated expansion of the spatial geometry of our brane-Universe, thus inducing a cosmological phase which is of inflationary type to all practical purposes.

This model has difficulties, however, if the space external to the branes is characterized by a geometry and a topology of the flat, Euclidean type: in that case, in fact, the interaction energy varies too rapidly with the spatial separation of the branes, and the induced inflationary phase does not meet the efficiency criteria needed to solve the problems of the standard scenario.

In order to obtain an efficient model we should be able to "slow down" the variation of the potential energy, by making it less sensitive to the distance between the two branes. This can be done in two ways.

A first possibility[24] relies on the assumption that the n extra dimensions, although geometrically flat, have the topology of a n-dimensional "torus", with spatial sections corresponding to circles of constant radius r.

When the separation of the brane–antibrane pair is of order of this radius, i.e., when $Y = r$, the toroidal topology of the space produces a n-dimensional "lattice" of effective images of the antibrane,[25] and the resulting potential energy acting

---

[24]See for instance the work by Burgess et al. [68].

[25]This optical lattice is due to the topology, because the gravitational force generated by the antibrane can act on the brane propagating along the circles of the torus in one direction an in

on the brane depends only on its displacement from the center of a "cell" of this lattice. In such a context, the variation of the potential turns out to be slow enough to guarantee the production of an efficient inflationary phase.

A second possibility[26] does not modify the topology of the extra-dimensional space, but introduces instead a non-trivial, curved geometry of anti-de Sitter type.[27]

In such a case, the spatial distance between our brane-Universe and the antibrane turns out to be deformed by the curved geometry, and the potential energy of their gravitational interactions is modified accordingly. The anti-de Sitter geometry, in particular, tends to "dilate" the spatial distance between the branes: as a consequence, the potential energy decreases more slowly with the distance, the evolution is slower, and the produced inflation meet the required efficiency criteria.

In summary, we can say that the cosmological models based on strings and membranes open up new and interesting prospects on the primordial evolution of our Universe. They provide us with different geometric interpretations of the inflationary dynamics (such as accelerated evolution of the initial perturbative vacuum, gravitational interaction between branes, etc.), and also possible mechanisms for the Big Bang explosion. But, above all, they suggest that the Universe could have existed—albeit in a form very different from the present one—even before the Big Bang.

Is there any hope of testing, in some direct or indirect experimental way, this last intriguing possibility?

6.3   Signals from Epochs Before the Big Bang?

If the Big bang does not correspond to an initial singularity (unlike what suggested by the standard cosmological model), and if the Universe, the space, the time, have existed even before the Big Bang (as suggested by string theory), the question arises whether some physical evidence of the primordial eras preceding the Big Bang has been produced, and has survived until the present epoch.

A possible clue to the existence of those primordial eras could be provided by the presence, in the cosmic background of electromagnetic radiation, of concentric circular regions characterized by temperature fluctuations much smaller than their typical average value.

---

the opposite one, thus reaching the brane from many points, as if many antibranes were present as sources of the total effective gravitational field.

[26]See for instance the paper by Kachru et al. [69].

[27]It is a geometry characterized by a negative cosmological constant, typically obtained as a possible solution of the gravitational equations describing the vacuum state of supersymmetric models (including, for instance, superstrings).

A signal of this type is predicted, for instance, by the so-called CCC (namely, Conformal Cyclic Cosmology) scenario.[28] According to this scenario, the evolution of our Universe is described by a continuous repetitions of cycles always more extended in space (and in time): each cycle starts with an initial Big Bang and ends with a phase of accelerated expansion like the one we are currently experiencing. There is no inflationary phase at the beginning of each cycle, as the final acceleration is enough to provide the whole amount of inflation required by the subsequent cycle.

In this scenario, unlike the ekpyrotic one, there are no thermodynamic restrictions on the possible number of allowed cycles. In each cycle, in fact, the dominant contribution to the total entropy comes from the production of gigantic "black holes" at the center of the galaxies,[29] each of them contributing with an entropy proportional to the area of its "event horizon." On the other hand, the cycles are so lasting in time that all the black holes have enough time to evaporate through a quantum process of radiation emission,[30] thus resetting the entropy to sufficiently low values before the beginning of the subsequent cycle.

Before completing the evaporation process, however, these gigantic black holes have plenty of time to collide among themselves. Their collisions are not so frequent but, when they occur, they release a huge amount of energy in the form of gravitational radiation, distributed over all frequency bands. This radiation is extremely penetrating, propagates throughout the duration of the cycle and is transmitted, in principle, also to the next cycle.

The primordial gravitational waves surviving the transition will appear, in the next cycle, as spherically symmetric bursts of energy centered around the points of space corresponding to the places of the collisions. This gravitational energy obviously interacts with all cosmic matter and radiation, affecting in particular the cosmic temperature in circular regions of space, and imprinting, as a mark, thermal fluctuations much lower than the average thermal variation of the other spatial regions.[31]

Given that a black hole may have experienced, during its life, more than one collision, it is possible that in the old cycle there have been several gravitational

---

[28]Described in a recent book by Penrose [70].

[29]A black hole is a concentrate of matter so dense as to be contained within a portion of space of radius smaller than its Schwarzschild radius r_s, a typical distance which for a static black hole is given by $r_s = 2ML_P^2$, where M is its total mass. If we consider the Earth, for instance, we find for the Schwarzschild radius a distance slightly less than 1 cm. For an observer outside the black hole, the surface of the Schwarzschild sphere of radius r_s represents the so-called event horizon: the gravitational attraction inside this surface is so strong that no classical object or signal is able to propagate to the outside, crossing the horizon.

[30]It is the so-called Hawking radiation. Taking into account quantum effects one finds, in fact, that the horizon of a static black hole behaves as a hot body radiating energy at a temperature which is inversely proportional to the horizon radius $2ML_P^2$. Because of this energy loss the mass of the black hole decreases in time and, consequently, the horizon radius becomes smaller and smaller, until it totally disappears at the end of the evaporation process.

[31]This effect is explained, for instance, in a recent paper by Gurzadyan e Penrose [71].

explosions, localized at about the same point in space but occurring at different times. This produces, in the new cycle, several spatial regions with a low thermal fluctuation which are (roughly) arranged as a series of concentric rings.

It is then amazing to discover that a series of anisotropic structures, with the shape of concentric rings, seems indeed to appear when performing appropriate analyses of the data on the cosmic radiation background [71]!

Unfortunately, it is not quite clear, so far, whether these "ring-shaped" anomalous regions should be really interpreted as physical signals from eras prior to "our" Big Bang, or should be included, instead, into the intrinsic "noise" (i.e., inaccuracy) of the current data. It seems, however, that if such ring anisotropies exist, and are not an artifact of the data analysis, they cannot be easily explained in the context of the standard cosmological model, even using its currently most accurate and updated version of ΛCDM type.

Fortunately enough there are other, even more direct, possibility to obtain evidence of the cosmological phases existed before the Big Bang, based on the observation of the stochastic background of relic gravitational radiation, and on the study of its spectrum.

6.3.1   The Cosmic Background of Relic Gravitons

As we have seen in Sect. 6.2, inflation can efficiently amplify the quantum fluctuations of the fields and of the geometry through a process which, from the point of view of quantum field theory, can be interpreted as the production of pairs of particles from the vacuum. The fluctuations of the space–time geometry, in particular, are associate to the fluctuations of the gravitational field: their amplification thus leads to the production of pairs of gravitons, and to the formation of a cosmic, stochastic background of relic gravitational waves.

Among all known types of radiation, on the other hand, the gravitational waves are by far the most penetrating ones. We can easily convince ourselves of this property by making a comparison with the electromagnetic waves.

In fact, the Universe became transparent to the electromagnetic radiation when the cosmic temperature dropped below the critical value of about 3,000 K (a value which is about a thousand times larger than the current temperature T_0). No doubt it is a very high temperature, typical of the Universe in a surely remote era. But it is nothing in comparison with the temperature below which the Universe became transparent to the gravitational radiation: the Planck temperature T_P, which is about 10^{32} times larger than the today temperature!

This means, in practice, that the cosmic gravitational waves may bring us "snapshots" of the early Universe dating back to epochs so remote as to have been "forgotten" forever by all other types of radiations and signals. The possible observation of a relic background of primordial gravitons would thus provide a unique tool to obtain first-hand information on the earliest, physically accessible, cosmological eras.

This could be the case, in particular, for a background of relic gravitational radiation reaching us from pre-big bang eras. If such a background exists, how could we recognize it and distinguish it, for instance, from the background of relic gravitons produced by a phase of standard inflationary evolution?

To answer this question we should first notice that the inflationary amplification of the gravitational fluctuations—that is to say, the production of pairs of graviton from the vacuum—does not occur with the same intensity for all quantum fluctuations. Depending on their wavelengths, certain fluctuations are amplified more efficiently than others (or less than others), and, as a consequence, the resulting "spectrum" of produced gravitons[32] is characterized by different intensities on different bands (or intervals) of frequency.

As already mentioned in Sect. 3.3, in fact, the quantum fluctuations can be decomposed into many tiny waves, oscillating with different frequencies. Such waves, just because of their nature of vacuum fluctuations, satisfy an important condition: their amplitude is proportional to the frequency, hence inversely proportional to the wavelength λ.

Because of the cosmological expansion, on the other hand, all frequencies tend to decrease in time, while (as already stressed in Sect. 3.2) the proper distances automatically increase with the dilation of the spatial geometry. The effective amplitude of the quantum fluctuations, which is proportional to the frequency, thus tends to decrease, while the wavelength λ grows in time.

The Hubble radius L_H, which controls the size of the cosmological horizon (i.e., of the spatial region within which all points have had enough time to exchange signals and interact) can also grow in time. During the inflationary phase, however, the growth of L_H is always slower[33] than the growth of λ. Hence, thanks to inflation, all wavelengths (even those which are initially very small) are doomed to become equal to the Hubble radius L_H, and later to exceed it.

From the time when λ exceeds L_H it makes no longer sense to talk of oscillations, as it is no longer possible to physically appreciate the variations in time and space of the intensity of those waves (as the separation of their maxima and minima is larger than the radius of the horizon). The amplitude of those quantum fluctuations thus stays "frozen" (i.e., constant, to all effects) at the value they had at the time when $\lambda = L_H$. Since the amplitude is proportional to $1/\lambda$, the frozen amplitude of each fluctuation turns out to be proportional to $1/L_H$, i.e., to the value of the curvature scale typical of the cosmic space–time at the time of freezing of the given fluctuation.

---

[32]With the name "spectrum", or spectral distribution, we shall precisely denote the mean energy of the produced gravitons per unit volume and per unit of logarithmic interval of frequency. This quantity represents, from a physical point of view, the energy density of the graviton background at any given fixed value of frequency.

[33]There are also inflationary models where the horizon radius L_H does not grow in time. For instance, it stays constant in models based on the de Sitter geometry, while it decreases in time in pre-big bang models based on string theory (see Sect. 6.1).

The various waves tend to "de-freeze" after inflation, when the Universe decelerates and L_H starts to grow faster than λ. What then happens is that all wavelengths, one after another, tend to become smaller than the Hubble radius: all quantum waves start to oscillate once again, and their amplitude restart decreasing, proportionally to their frequency. The freezing phase, however, has stopped the decrease of the amplitude for a long period, and has thus produced an effective amplification of all fluctuations.

The final amplitude of a wave, after such an amplification process, depends on its freezing value which, in turn, depends on the evolution of the Hubble radius L_H during the phase of inflation. It should be noticed, in this respect, that different waves come to satisfy the freezing condition $\lambda = L_H$ at different epochs: in fact, the smaller the value of the initial wavelength, the longer the time needed for λ to grow enough, and to come to meet the condition $\lambda = L_H$.

As a consequence, the freezing amplitude will be the same for all waves only if L_H is constant in time during the whole inflationary phase. If L_H is increasing then short wavelengths will freeze later, and will have a freezing amplitude (which is proportional to $1/L_H$) smaller than the freezing amplitude of long wavelengths. If L_H decreases in time, on the contrary, exactly the opposite will happen.

But, as already noted, $1/L_H$ is also proportional to the space–time curvature. If the curvature decreases during the inflationary phase then the shortest wavelengths will be amplified less than the others; if the curvature increases, instead, the shortest wavelengths will be amplified more than the others.

The wavelength λ, on the other hand, is inversely proportional to the frequency (and to the energy) of the wave: small wavelengths correspond to large frequencies, and vice versa. We can thus summarize the previous arguments by saying that, in general, the inflationary amplification of the quantum fluctuations as a function of the frequency tends to follow the behavior in time of the curvature during the inflationary phase. This means that the spectral intensity of the produced gravitational radiation is constant, growing or decreasing with frequency depending on if the inflationary phase has a curvature which is constant, growing or decreasing in time.[34]

Thanks to this simple—but important—result we are able to distinguish, in principle, the relic gravitons produced by a phase of standard inflation, at decreasing curvature, from the relic gravitons produced by a phase of pre-big bang inflation, at growing curvature, typical of string cosmology models[35] (see for instance the Figs. 6.3 and 6.4). If we could ever detect a cosmic background of relic gravitons, measure its spectrum, and determine if it is increasing (or decreasing) with frequency, then we would know if it was produced before (or after) the Big Bang.

---

[34]Growing spectra are also called "blue", while decreasing spectra are called "red".

[35]The presence of a relic graviton background more intense at high frequencies than at low frequencies as a typical signature of an inflationary phase occurring *before* the Big Bang was suggested by Gasperini and Giovannini [72], and later further studied by Brustein, Gasperini, and Veneziano [73].

Until now, unfortunately, no primordial background of relic gravitational radiation has ever been observed.[36]

This is something quite understandable, as such a background, if existing, would be characterized by an extremely weak intensity: in the point of maximum intensity, its present energy density could be not larger than about 1 millionth of the (already very small) energy density of cosmic dark matter.[37] We should also recall that gravity, among all fundamental interactions, is by far the weakest one, and the most difficult to be experimentally studied.

Nevertheless, there are in principle various possibilities of detecting a cosmic background of relic gravitons, with both direct and indirect methods. For our purposes it will be enough to recall two of them: a direct detection through the currently existing gravitational antennas,[38] and an indirect one through the possible effects of the relic gravitons on the cosmic background of electromagnetic radiation.

Gravity is indeed a universal interaction, and the gravitational waves are coupled to all forms of matter and energy. The relic gravitons produced in the primordial Universe can interact, in particular, with the photons of the cosmic radiation background, and affect their physical properties.

The relic gravitons, for instance, can imprint on the temperature (and density) of the cosmic radiation characteristic variations with the shape of rings centered around a point, if—as discussed in the previous section—they have been emitted by primordial energy explosions. If the relic gravitons have been generated by inflation, instead, they modify the photon temperature producing anisotropies and inhomogeneities stochastically distributed among all points in space.

The gravitons of inflationary origin can also affect in a peculiar and characteristic way the so-called "polarization" of the cosmic electromagnetic radiation, namely the direction along which the oscillations of the electromagnetic waves are oriented. In particular, the presence of a stochastic background of relic gravitons can induce in the cosmic photons a typical polarization state called "B mode" (or magnetic mode), characterized by the fact that the polarization directions tend to be arranged in the form of vortices around the warmest points and the coldest points.

It is important to stress that the temperature anisotropies and inhomogeneities can also be produced by relic radiation of scalar type, while the production of the B-mode polarization states is an effect typically due to the presence of a relic

---

[36]See, however, the NOTE ADDED IN PROOF at the end of this chapter.

[37]Were it larger, in fact, it would have affected the cosmological dynamics since the nucleosynthesis epoch, in contrast with the results of current observations.

[38]They are instruments able to respond to the passage of a gravitational wave, with the function of amplifying the electromechanical effects produced by the wave, and providing a signal strong enough to be detectable. The available gravitational antennas are currently of two types, based, respectively, on the mechanism of the resonant mechanical bar and of the interferometer. There are projects, already in advanced phase, of interferometric antennas to be launched in space and put in orbit around the Sun, in order to achieve a better sensitivity in the low-frequency bands of the gravitational spectrum (see e.g., the book by Maggiore [74]; see also [42] for a discussion focalized on the detection of relic gravitons, or [3] for a textbook in Italian).

graviton background. A possible observation of this polarization mode would thus provide an important—though indirect—experimental evidence of the existence of such a background.

Given the present experimental sensitivities, a positive result in the detection of the B-mode states would imply that the graviton background is strong enough at the very low frequencies at which the anisotropies and the polarizations of the cosmic electromagnetic radiation are currently studied and observed.[39] In these frequency bands, in order to produce a currently detectable signal, the energy density of the graviton background should be not much smaller than about one ten-billionth (i.e., 10^{-10}) of the dark-matter density.

A direct detection of the relic graviton background would imply, instead, that the background is strong at high enough frequencies. The present antennas (including those not yet operating, but planned to be operative in a near future) are sensitive, in fact, to the gravitational waves in a spectrum of frequencies ranging from the millihertz (10^{-3} Hz) to the kilohertz (10^3 Hz) band.

In these frequency bands, the minimum detectable intensity depends on the type of antenna: currently, to be detectable, the graviton background is required to have an energy density of at least one hundred thousandth (i.e., 10^{-5}) of the dark matter density. Such a limiting value should be lowered down to 10^{-10}–10^{-11}, in a near future, mainly thanks to the antennas operating in space.

The direct observation of the relic gravitons produced by inflation, or the observation of the effects induced by the relic gravitons on the polarization states of the cosmic photons, represents a difficult but exciting challenge to experimental physics for the coming years. However things go, it is encouraging to note that the situation, at least from a theoretical point of view, is conceptually quite simple.

If we were to obtain positive results from both types of experiment, then we could compare the intensity of the relic gravitational radiation at low and high frequency, determine the behavior (growing or decreasing with frequency) of the spectrum, and thus know whether the gravitons have been produced before or after the epoch of the Big Bang.

If we were to observe the B-mode polarization effects at low frequencies, without directly detecting, however, the graviton background in the sensitivity range of the present antennas, the result would not be conclusive (at least until the antennas do not reach the sensitivity level of the polarization measurements).

Nevertheless, we would obtain a strong indication that the spectrum is decreasing,[40] and that the gravitons have been produced by a phase of standard inflation: in fact, the spectra of growing type tend to be quite "steep," thus providing a negligible contribution to the polarization effects at low frequencies (as illustrated in Fig. 6.7).

---

[39]The anisotropies and inhomogeneities of the cosmic electromagnetic radiation are currently measured with the highest precision at angular scales of the order of one degree (or slightly lower), corresponding to the fluctuations of wavelengths ranging from about $\lambda = L_H$ to $\lambda = 0.01 L_H$, where L_H is the Hubble radius. The corresponding frequencies range from 10^{-18} to 10^{-16} Hz.

[40]See, however, the NOTE ADDED IN PROOF at the end of this chapter.

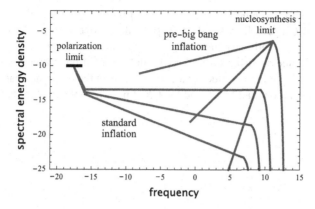

Fig. 6.7 Possible behavior of the energy density of a cosmic background of relic gravitons as a function of frequency. The energy spectrum is constant or decreasing for gravitons produced by a phase of standard inflation, while it is growing for gravitons produced by an inflationary phase of pre-big bang type, typical of string cosmology models. The figure illustrates the case in which the decreasing spectra are bounded by the maximum value allowed by current observations of the polarization of the cosmic radiation, while the growing spectra are bounded by the maximum value allowed by current nucleosynthesis data. See for instance the textbooks [2, 42] for a detailed discussion

On the contrary, if we were to detect the graviton background and observe, at high frequency, a level of energy density higher than the one required to produce the polarization effects typical of the low-frequency limit, we would obtain a strong indication that the spectrum is growing, and that the gravitons have been produced by a phase of pre-big bang inflation, as predicted by string cosmology (see Fig. 6.7).

For the moment we can just wait, hoping that Nature, as usual, sooner or later will repay our efforts and our desire for knowledge.

Note Added in Proof

Just a few days before receiving the proofs of this book there was the announcement, on 17 March 2014, that the polarization properties of the CMB radiation measured by the BICEP2 experiment seem to exactly follow the trend we should expect if such a polarization would be produced in the presence of a relic background of primordial gravitational radiation (see the previous discussion of the B-mode polarization state). At the time of writing this note (today, 5 April 2014), the intensity and the spectral tilt of the (possibly detected) background of cosmic gravitons is still uncertain. There are indication, however, that the spectrum could be "blue", i.e., growing with frequency (see, e.g., the paper by M. Gerbino et al., "Blue gravity waves from BICEP2?" [arXiv:14035732]). If this property will be confirmed, it would provide an (indirect) indication that the blue spectrum of pre-big bang gravitons, predicted by string-cosmology models of inflation, grows with frequency only slightly and is able to produce detectable effects in the frequency bands relevant to both the gravitational antennas and the CMB polarization (see Fig. 6.7).

Chapter 7
Conclusion

In the pages of this book we have traveled in space and time, for very small and very large distances, sometimes in the past and sometimes in the future. We have introduced things that we do not know exactly how to explain, and we have provided explanations and models for things that probably we will never experience.

But, above all, we have tried to imagine the beauty and the symmetry hidden in the unity of the most different forms of matter and energy, and in the simplicity of the most complicated physical processes.

We have also realized that maybe there are no limits to our level of understanding the physical reality in which we are immersed, because, whenever we believe we have built an efficient, complete, and ultimate model, there are to appear new interactions, new spatial dimensions, new fundamental objects, new physical effects, new cosmological eras ...

My conclusion, at this point, is very simple. Before finding a satisfactory solution for all the problems that physics currently poses to our attention (for instance, what is the theory able to describe all the fundamental forces of Nature, how many are the dimensions of space, how was born our Universe, how it evolved in time, and what will be its future evolution, etc.), I am sure we have still in front of us many years of work and—definitely—of surprising findings.

M. Gasperini, *Gravity, Strings and Particles*, DOI 10.1007/978-3-319-00599-7_7,
© Springer International Publishing Switzerland 2014

References

1. R. Durrer, *The Cosmic Microwave Background* (Cambridge University Press, Cambridge, 2008)
2. S. Weinberg, *Cosmology* (Oxford University Press, Oxford, 2008)
3. M. Gasperini, *Lezioni di Cosmologia Teorica* (Spriger, Milano, 2012)
4. E.G. Adelberger, B.R. Heckel, A.E. Nelson, Ann. Rev. Nucl. Part. Sci. **53**, 77 (2003)
5. E. Fischbach, D. Sudarsky, A. Szafer, C. Talmadge, S.H. Aronson, Phys. Rev. Lett. **56**, 3 (1986)
6. R. Barbieri, S. Cecotti, Z. Phys. **C 33**, 255 (1986)
7. M. Gasperini, Phys. Rev. **D 40**, 325 (1989)
8. M. Gasperini, Phys. Rev. **D 63**, 047301 (2001)
9. T. Taylor, G. Veneziano, Phys. Lett. **B 213**, 459 (1988)
10. M. Gasperini, Phys. Lett. **B 327**, 314 (1994);
    M. Gasperini, G. Veneziano, Phys. Rev. **D 50**, 2519 (1994)
11. J. Khoury, A. Weltman, Phys. Rev. **D 69**, 044026 (2004)
12. R. Sundrum, Phys. Rev. **D 69**, 044014 (2004)
13. T. Kaluza, Sitzungsber. Preuss. Akad. Wiss. Berlin **1921**, 966 (1921);
    O. Klein, Z. Phys. **37**, 895 (1926)
14. N. Arkani Hamed, S. Dimopoulos, G.R. Dvali, Phys. Lett. **B 429**, 263 (1998)
15. I. Antoniadis, Phys. Lett. **B 246**, 377 (1990)
16. L. Randall, R. Sundrum, Phys. Rev. Lett. **83**, 4960 (1999)
17. A.G. Riess et al., Astron. J. **116**, 1009 (1998);
    S. Perlmutter et al., Astrophys. J **517**, 565 (1999)
18. G. Dvali, G. Gabadadze, M. Porrati, Phys. Lett. **B 485**, 208 (2000)
19. I. Slatev, L.Wang, P.J. Steinhardt, Phys. Rev. Lett. **82**, 869 (1999)
20. C. Armendariz-Picon, V. Mukhanov, P.J. Steinhardt, Phys. Rev. Lett. **85**, 4438 (2000)
21. M. Gasperini, Phys. Rev. **D 64**, 043510 (2001)
22. M. Gasperini, F. Piazza, G. Veneziano, Phys. Rev. **D 65**, 023508 (2001)
23. L. Amendola, M. Gasperini, F. Piazza, JCAP **09**, 014 (2004); Phys. Rev. **D 4**, 127302 (2006)
24. S. Weinberg, Rev. Mod. Phys. **61**, 1 (1989)
25. B. Zumino, Nucl. Phys. **B 89**, 535 (1975)
26. E. Witten, in *Sources and Detection of Dark Matter and Dark Energy in the Universe*, ed. by D.B. Cline (Springer, Berlin, 2001), p. 27
27. R. Bousso, Gen. Rel. Grav. **40**, 607 (2008)
28. T. Padmanabhan, Gen. Rel. Grav. **40**, 529 (2008)
29. A.Kobakhidze, N.L. Rodd, Int. J. Theor. Phys. **52**, 2636 (2013)
30. M. Gasperini, JHEP **06**, 009 (2008)
31. I. Antoniadis, E. Dudas, A. Sagnotti, Phys. Lett. **B 464**, 38 (1999)

M. Gasperini, *Gravity, Strings and Particles*, DOI 10.1007/978-3-319-00599-7,
© Springer International Publishing Switzerland 2014

32. G.F.R. Ellis, *Spacetime and the Passage of Time*, arXiv:1208.2611, published in "Springer Hanbook of Spacetime", ed. by A. Ashtekar, V. Petkov (Springer, Berlin, 2014)
33. P.C.W. Davies, Sci. Am. (Special Edition) **21**, 8 (2012)
34. J.B. Barbour, *The End of Time: The Next Revolution in Physics* (Oxford University Press, Oxford, 1999)
35. J.D. Barrow, D.J. Shaw, Phys. Rev. Lett. **106**, 101302 (2011)
36. N. Kaloper, K.A. Olive, Phys. Rev. **D 57**, 811 (1998)
37. E. Caianiello, La Rivista del Nuovo Cimento **15**, 1 (1992)
38. M. Gasperini, Int. J. Mod. Phys. **D 13**, 2267 (2004)
39. B. Zwiebach, *A First Course in String Theory* (Cambridge University Press, Cambridge, 2009)
40. M.B. Green, J. Schwartz, E. Witten, *Superstring Theory* (Cambridge University Press, Cambridge, 1987)
41. J. Polchinski, *String Theory* (Cambridge University Press, Cambridge, 1998)
42. M. Gasperini, *Elements of String Cosmology* (Cambridge University Press, Cambridge, 2007)
43. M.A. Virasoro, Phys. Rev. **D 1**, 2933 (1970)
44. P. Ramond, Phys. Rev. **D 3**, 2415 (1971)
45. A. Neveau, J.H. Schwarz, Nucl. Phys. **B 31**, 86 (1971)
46. F. Gliozzi, J. Scherk, A. Olive, Nucl. Phys. **B 122**, 253 (1977)
47. A. Sagnotti, J. Phys. **A 46**, 214006 (2013)
48. K. Kikkawa, M.Y. Yamasaki, Phys. Lett. **B 149**, 357 (1984)
49. N. Sakai, I. Senda, Prog. Theor. Phys. **75**, 692 (1984)
50. A.A. Tseytlin, Mod. Phys. Lett. **A 6**, 1721 (1991)
51. G. Veneziano, Phys. Lett. **B 265**, 287 (1991)
52. E. Witten, Nucl. Phys. **B 443**, 85 (1995)
53. R. Bousso, J. Polchinski, Sci. Am. **291**, 60 (2004)
54. R. Brandenberger, C. Vafa, Nucl. Phys. **B 316**, 391 (1989)
55. A.A. Tseytlin, C. Vafa, Nucl. Phys. **B 372**, 443 (1992)
56. S. Alexander, R. Brandenberger, D. Easson, Phys. Rev. **D 62**, 103509 (2000)
57. S. Weinberg, *Gravitation and Cosmology* (Wiley, New York, 1972)
58. A.A. Penzias, R.W. Wilson, Astrophys. J. **142**, 419 (1965)
59. J. Smooth et al., Astrophys. J. **396**, L1 (1992)
60. A. Guth, Phys. Rev. **D 23**, 347 (1981)
61. E.W. Kolb, M.S. Turner, *The Early Universe* (Addison Wesley, Redwood City, 1990)
62. A. Borde, A. Guth, A. Vilenkin, Phys. Rev. Lett. **90**, 151301 (2003)
63. M. Gasperini, G. Veneziano, Astropart. Phys. **1**, 317 (1993); Phys. Rep. **373**, 1 (2003)
64. M. Gasperini, *The Universe Before the Big Bang: Cosmology and String Theory* (Springer, Berlin, 2008)
65. J. Khoury, B.A. Ovrut, P.J. Steinhardt, N. Turok, Phys. Rev. **D 64**, 123522 (2001)
66. P.J. Steinhardt, N. Turok, Phys. Rev. **D 65**, 126003 (2002)
67. N. Goheer, M. Kleban, L. Susskind, JHEP **0307**, 056 (2003)
68. C. Burgess et al., JHEP **0107**, 047 (2001)
69. S. Kachru et al., JCAP **0310**, 013 (2003)
70. R. Penrose, *Cycles of Time: An Extraordinary New View of the Universe* (Bodley Head, London, 2010)
71. V.G. Gurzadyan, R. Penrose, Eur. Phys. J. Plus **128**, 22 (2013)
72. M. Gasperini, M. Giovannini, Phys. Lett. **B 282**, 36 (1992); Phys. Rev. **D 47**, 1519 (1993)
73. R. Brustein, M. Gasperini, G. Venenziano, Phys. Lett. **B 361**, 45 (1995)
74. M. Maggiore, *Gravitational Waves* (Oxford University Press, Oxford, 2007)

Index

M. Gasperini, *Gravity, Strings and Particles*, DOI 10.1007/978-3-319-00599-7,
© Springer International Publishing Switzerland 2014